Lecture Notes in Applied and Computational Mechanics

Volume 62

Series Editors

Friedrich Pfeiffer
Peter Wriggers

For further volumes:
www.springer.com/series/4623

Lecture Notes in Applied and Computational Mechanics

Edited by F. Pfeiffer and P. Wriggers

Further volumes of this series found on our homepage: springer.com

Vol. 61: Frémond, M., Maceri, F., (Ed.)
Mechanics, Models and Methods in Civil Engineering
498 p. 2012 [978-3-642-24637-1]

Vol. 59: Markert, B., (Ed.)
Advances in Extended and Multifield Theories for Continua
219 p. 2011 [978-3-642-22737-0]

Vol. 58: Zavarise, G., Wriggers, P. (Eds.)
Trends in Computational Contact Mechanics
354 p. 2011 [978-3-642-22166-8]

Vol. 57: Stephan, E., Wriggers, P.
Modelling, Simulation and Software Concepts for Scientific-Technological Problems
251 p. 2011 [978-3-642-20489-0]

Vol. 54: Sanchez-Palencia, E., Millet, O., Béchet, F.
Singular Problems in Shell Theory
265 p. 2010 [978-3-642-13814-0]

Vol. 53: Litewka, P.
Finite Element Analysis of Beam-to-Beam Contact
159 p. 2010 [978-3-642-12939-1]

Vol. 52: Pilipchuk, V.N.
Nonlinear Dynamics: Between Linear and Impact Limits
364 p. 2010 [978-3-642-12798-4]

Vol. 51: Besdo, D., Heimann, B., Klüppel, M., Kröger, M., Wriggers, P., Nackenhorst, U.
Elastomere Friction
249 p. 2010 [978-3-642-10656-9]

Vol. 50: Ganghoffer, J.-F., Pastrone, F. (Eds.)
Mechanics of Microstructured Solids 2
102 p. 2010 [978-3-642-05170-8]

Vol. 49: Hazra, S.B.
Large-Scale PDE-Constrained Optimization in Applications
224 p. 2010 [978-3-642-01501-4]

Vol. 48: Su, Z.; Ye, L.
Identification of Damage Using Lamb Waves
346 p. 2009 [978-1-84882-783-7]

Vol. 47: Studer, C.
Numerics of Unilateral Contacts and Friction
191 p. 2009 [978-3-642-01099-6]

Vol. 46: Ganghoffer, J.-F., Pastrone, F. (Eds.)
Mechanics of Microstructured Solids
136 p. 2009 [978-3-642-00910-5]

Vol. 45: Shevchuk, I.V.
Convective Heat and Mass Transfer in Rotating Disk Systems
300 p. 2009 [978-3-642-00717-0]

Vol. 44: Ibrahim R.A., Babitsky, V.I., Okuma, M. (Eds.)
Vibro-Impact Dynamics of Ocean Systems and Related Problems
280 p. 2009 [978-3-642-00628-9]

Vol.43: Ibrahim, R.A.
Vibro-Impact Dynamics
312 p. 2009 [978-3-642-00274-8]

Vol. 42: Hashiguchi, K.
Elastoplasticity Theory
432 p. 2009 [978-3-642-00272-4]

Vol. 41: Browand, F., Ross, J., McCallen, R. (Eds.)
Aerodynamics of Heavy Vehicles II: Trucks, Buses, and Trains
486 p. 2009 [978-3-540-85069-4]

Vol. 40: Pfeiffer, F.
Mechanical System Dynamics
578 p. 2008 [978-3-540-79435-6]

Vol. 39: Lucchesi, M., Padovani, C., Pasquinelli, G., Zani, N.
Masonry Constructions: Mechanical Models and Numerical Applications
176 p. 2008 [978-3-540-79110-2]

Vol. 38: Marynowski, K.
Dynamics of the Axially Moving Orthotropic Web
140 p. 2008 [978-3-540-78988-8]

Vol. 37: Chaudhary, H., Saha, S.K.
Dynamics and Balancing of Multibody Systems
200 p. 2008 [978-3-540-78178-3]

Vol. 36: Leine, R.I.; van de Wouw, N.
Stability and Convergence of Mechanical Systems with Unilateral Constraints
250 p. 2008 [978-3-540-76974-3]

Vol. 35: Acary, V.; Brogliato, B.
Numerical Methods for Nonsmooth Dynamical Systems: Applications in Mechanics and Electronics
545 p. 2008 [978-3-540-75391-9]

Vol. 34: Flores, P.; Ambrósio, J.; Pimenta Claro, J.C.; Lankarani Hamid M.
Kinematics and Dynamics of Multibody Systems with Imperfect Joints: Models and Case Studies
186 p. 2008 [978-3-540-74359-0]

Alan T. Zehnder

Fracture Mechanics

Springer

Alan T. Zehnder
Sibley School of Mechanical and
Aerospace Engineering
Cornell University
Ithaca
USA
atz2@cornell.edu

ISSN 1613-7736 e-ISSN 1860-0816
Lecture Notes in Applied and Computational Mechanics
ISBN 978-94-007-2594-2 e-ISBN 978-94-007-2595-9
DOI 10.1007/978-94-007-2595-9
Springer London Dordrecht Heidelberg New York

British Library Cataloguing in Publication Data
A catalogue record for this book is available from the British Library

Library of Congress Control Number: 2011944214

Printed on acid-free paper

Springer is part of Springer Science+Business Media (www.springer.com)

Preface

Fracture mechanics is a large and always growing field. A search of the Cornell Library in winter 2006 uncovered over 181 entries containing "fracture mechanics" in the subject heading and 10,000 entries in a relevance keyword search. This book is written for students who want to begin to understand, apply and contribute to this important field. It is assumed that the reader is familiar with the theory of linear elasticity, vector calculus, linear algebra and indicial notation.

There are many approaches to teaching fracture. Here the emphasis is on continuum mechanics models for crack tip fields and energy flows. A brief discussion of computational fracture, fracture toughness testing and fracture criteria is given. They contain very little on fracture at the micromechanical level or on applications. Both the mechanics and the materials sides of fracture should be studied in order to obtain a balanced, complete picture of the field. So, if you start with fracture *mechanics*, keep going, study the physical aspects of fracture across a broad class of materials and read up on fracture case studies [1] to learn about applications.

I use these notes in a one-semester graduate level course at Cornell. Although these notes grow out of my experience teaching, they also owe much to Ares Rosakis from whom I took fracture mechanics at Caltech and to Hutchinson's notes on nonlinear fracture [2]. Textbooks consulted include Lawn's book on the fracture of brittle materials [3], Suresh on fatigue [4] and Janssen [5], Anderson [6], Sanford [7], Hellan [8] and Broberg [9].

I would like to thank Prof. E.K. Tschegg for generously hosting me during my 2004 sabbatical leave in Vienna, during which I started these notes. Thanks also to my students who encouraged me to write and, in particular, to former students Mike Czabaj and Jake Hochhalter who each contributed sections.

References

1. ASTM, *Case Histories Involving Fatigue and Fracture, STP 918* (ASTM International, West Cohshohocken, 1986)
2. J.W. Hutchinson, *A Course on Nonlinear Fracture Mechanics* (Department of Solid Mechanics, The Technical University of Denmark, 1979)

3. B. Lawn, *Fracture of Brittle Solids*, 2nd edn. (Cambridge University Press, Cambridge, 1993)
4. S. Suresh, *Fatigue of Materials*, 2nd edn. (Cambridge University Press, Cambridge, 1998)
5. M. Janssen, J. Zuidema, R. Wanhill, *Fracture Mechanics*, 2nd edn. (Spon Press, London, 2004)
6. T.L. Anderson, *Fracture Mechanics Fundamentals and Applications*, 2nd edn. (CRC Press, Boca Raton, 1995)
7. R.J. Sanford, *Principles of Fracture Mechanics* (Prentice Hall, New York, 2003)
8. K. Hellan, *Introduction to Fracture Mechanics* (McGraw-Hill, New York, 1984)
9. K.B. Broberg, *Cracks and Fracture* (Academic Press, San Diego, 1999)

Ithaca, USA

Alan T. Zehnder

Contents

Acronyms

a	crack length
$\dot{a}$	crack growth rate
b	uncracked ligament length
b_n	series coefficients
a_λ, b_λ	constants in displacement potential
$\mathbf{b}$	body force
c	fracture sample dimension
d_n	numerical factor in CTOD relationship
$\mathbf{e}_i$	basis vectors
f, g	functions
f_σ	stress optic coefficient
f	yield surface
$\tilde{f}$	angular variation in HRR field
h	fracture sample dimension
$h(z)$	analytic function in anti-plane shear theory
$k_I,\ k_{II}$	stress intensity factors before kinking
k_n	plastic stress intensity
l	kinked crack length
$\mathbf{m}$	normal vector
m	summation integer
m_0	initial slope of load displacement curve in chevron notch test
n	power law exponent, or summation integer
$\tilde{n}$	power law exponent
$n(z) = n_1 + in_2$	complex normal
$\mathbf{n}$	normal vector
$\mathbf{p}$	crack face traction
$\mathbf{p}_i$	principal directions
q	generalized displacement
$\tilde{q}$	function in domain integral
r	distance from crack tip, or radial coordinate
r_c	compliance ratio in chevron notch fracture analysis

r_p	plastic zone size
s	fracture surface, or yield zone length in Dugdale model
t	time or dummy integration variable
$\mathbf{t}$	traction
$\mathbf{u}$	displacement
u	x component of displacement
v	y component of displacement
v_L	load-line displacement
v_C	crack mouth opening displacement
w	x_3 component of displacement
$\mathbf{x}$	coordinates
x, y	coordinates
$z = x_1 + ix_2$	complex coordinate
A	a constant
$A_\lambda, B_\lambda, C_\lambda, D_\lambda$	constants in stress function
B	plate thickness
C	compliance, or a constant
$\tilde{C}$	cross correlation coefficient
D	bending stiffness
E	elastic modulus
E'	E (plane stress), $E/(1-\nu^2)$ (plane strain)
F, F_1, F_2	functions, used in various places
$\mathbf{F}$	nodal force
G	energy release rate
G_c	fracture toughness in terms of G
H_T	true hardness
H_V	Vicker's hardness
H	mapped element y dimension, or hardening slope
H_{ij}, G_{ij}	functions in moving crack analysis
$I, I', \tilde{I}, \tilde{I}'$	intensity patterns
I_n	numerical factor in HRR field
J	J integral
J_2	second invariant of deviatoric stress
K_I, K_{II}, K_{III}	stress intensity factors
K_{IC}, K_{IVM}, K_{wof}	fracture toughness values
K_Q	provisional fracture toughness
KE	kinetic energy density
L	mapped element x dimension
N	cycle count, or fringe order
N_i	shape functions
P	generalized force
p_i, P_i, q_i, Q_i	locations in DIC method
Q	factor in J–Q theory
R	load ratio in cyclic loading, or dimension of a circular region, or a constant in DCB test method

$R(r), T(\theta)$	r and θ separable functions
$\mathbf{S}$	deviatoric stress
S	sample dimension
T	T stress
$T,\ L$	transverse and rolling directions
T_0	tearing modulus
U	work
W	strain energy density, or a width dimension for fracture samples
Y	non-dimensional stress intensity factor in chevron notch test
$\mathscr{A}$	area
$\mathscr{C}$	contour
$\mathscr{S}$	surface
$\mathscr{S}_t$	surface over which traction prescribed
$\mathscr{S}_u$	surface over which displacement prescribed
$\mathscr{V}$	volume
α	non-dimensional parameter, hardening parameter, or strain gauge angle
β	load angle
α, β	lines of maximum shear stress
γ	strain
γ_α	$2\gamma_{3\alpha}$
γ_0	yield strain
δ	displacement across crack in cohesive zone
δ_T	crack tip opening displacement
δ_C	critical crack opening displacement
η, ξ	finite element parent coordinates
η	normalized T stress
θ	angle, angular coordinate
θ_0	crack perturbation angle
κ	$(3-4\nu)/(1+\nu)$ (plane stress), $(3-4\nu)$ (plane strain)
λ	eigenvalue or total deformation magnitude
λ^p	plastic flow magnitude
$\lambda(x)$	crack path
λ_m	ductility parameter
μ	shear modulus
ν	Poisson's ratio
$\bar{\nu}$	$\nu/(1+\nu)$ (plane stress), ν (plane strain)
$\chi(z)$	stress function in anti-plane shear theory
ρ	density, or inner zone radius in SSY model
σ	stress
σ_∞	far field tension stress
σ_0	yield stress in tension
σ_{TS}	ultimate tensile strength
σ_Y	$(\sigma_0+\sigma_{TS})/2$

$\hat{\sigma}$	cohesive strength
τ_α	$\sigma_{3\alpha}$
$\tau(z)$	$\tau_1 + i\tau_2$, complex shear stress
τ_∞	far field out-of-plane shear stress
τ_0	yield stress in shear
$\phi(z)$	analytic function in Westergaard solution
φ	angle between coordinate axes and slip lines
$\psi(z)$	analytic function in Westergaard solution
ψ_C	crack tip opening angle
Γ	contour in 2D
Γ_0	cohesive zone energy
Π	potential energy
Φ	stress function
Ψ	displacement potential
Ω	total strain energy

Chapter 1
Introduction

Abstract The consequences of fracture can be minor or they can be costly, deadly or both. Fracture mechanics poses and finds answers to questions related to designing components and processes against fracture. The driving forces in fracture mechanics are the loads at the crack tip, expressed in terms of the stress intensity factor and the energy available to the crack tip. The resistance of the material to fracture is expressed in terms of fracture toughness. Criteria for fracture can be stated as a balance of the crack tip loads and the material's fracture resistance, or "toughness."

1.1 Notable Fractures

Things break everyday. This you know already. Usually a fracture is annoying and perhaps a little costly to deal with, a broken toy, or a cracked automobile windshield. However, fractures can also be deadly and involve enormous expense.

The deHavilland Comet, placed in service in 1952, was the world's first jetliner [1]. Pressurized and flying at high speed and altitude, the Comet cut 4 hours from the New York to London trip. Tragically two Comets disintegrated in flight in January and April 1954 killing dozens. Tests and studies of fragments of the second of the crashed jetliners showed that a crack had developed due to metal fatigue near the radio direction finding aerial window, situated in the front of the cabin roof. This crack eventually grew into the window, effectively creating a very large crack that failed rapidly, leading to the crashes. A great deal was learned in the investigations that followed these incidents [2, 3] and the Comet was redesigned to be structurally more robust. However, in the four years required for the Comet to be re-certified for flight, Boeing released its 707 taking the lead in the market for jet transports.

However, Boeing was not to be spared from fatigue fracture problems. In 1988 the roof of the forward cabin of a 737 tore away during flight, killing a flight attendant and injuring many passengers. The cause was multiple fatigue cracks linking up to form a large, catastrophic crack [4, 5]. The multitude of cycles accumulated on this aircraft, corrosion and maintenance problems all played a role in this accident. Furthermore, the accident challenged the notion that fracture was well understood and under control in modern structures.

This understanding was again challenged on 17 November 1994, 4:31am PST, when a magnitude 6.7 earthquake shook the Northridge Valley in Southern Califor-

A.T. Zehnder, *Fracture Mechanics*,
Lecture Notes in Applied and Computational Mechanics 62,
DOI 10.1007/978-94-007-2595-9_1,

Fig. 1.1 Fracture surface of broken ICE train wheel tire. Reprinted from Engineering Failure Analysis, Vol 11, V. Esslinger, R. Kieselbach, R. Koller, and B. Weisse, "The railway accident of Eschede—technical background," 515–535, copyright (2004), with permission from Elsevier [7]

nia for 15 seconds. The damage was severe: 57 people lost their lives, 1500 were injured and 12,500 buildings were damaged. That damage occurred is no surprise, however, what did surprise structural engineers were the fractures in many welded beam-column joints in steel framed buildings. These joints, designed to absorb energy by plastic deformation, instead fractured in an almost brittle fashion [6]. Due to such fractures over 150 buildings were damaged. In one the damage was so severe the building was demolished; others had to be evacuated.

The German Intercityexpress, or ICE, offers high speed, comfortable train travel at speeds up to 280 km/hr. On 5 June 1998 ICE 884, traveling on the Munich-Hamburg route at a speed of 250 km/hr crashed near the village of Eschede resulting in 100 deaths, 100 injuries, the destruction of a bridge, the track, the train and interruption of train service. The cause and course of the accident are described by Esslinger et al. [7]:

> The tire detached from the wheel, was dragged along, jammed under the floor of the carriage and then got stuck in the tongue of a switch. By this the switch was toggled to the neighboring track and the hind part of the train redirected there. This led to derailment and collision of the derailed train part with the pylon of a road bridge leading over the tracks. The collapsing bridge buried a part of the train.

The cause of the tire detachment was a fatigue crack, see Fig. 1.1 that grew from the inner rim of the tire. The crack grew slowly by fatigue to about 80% of the cross sectional area of the tire before the final, rapid fracture.

On 2 November 2007 a Missouri Air National Guard F-15C broke in two in flight. The pilot ejected but sustained injuries. Subsequent investigations revealed a manufacturing defect in which a fuselage longeron was machined to below its design thickness. The thinned longeron stressed to higher than planned levels, failed by initiation and growth of a fatigue crack that grew to a critical length before final, rapid fracture [8]. The entire US Air Force fleet of F15s was, at a time of war, grounded for some time following the 2 November accident. Newer F15Es were quickly returned to flight but older F15A-D models were returned to flight only after inspection of each vehicle, over 180 of which showed the manufacturing defect and 9 of which contained similar longeron cracks. Repair was estimated at $250,000 per vehicle. The repair costs and large number of aircraft with the same defect called into question the continued use of the F15A-D fleet.

1.2 Basic Fracture Mechanics Concepts

It should be clear that fracture is a significant problem in the industrialized world and that a theoretical and practical basis for design against fracture is needed. Fracture mechanics deals essentially with the following questions: Given a structure or machine component with a preexisting crack or crack-like flaw what loads can the structure take as a function of the crack size, configuration and time? Given a load and environmental history how fast and in what directions will a crack grow in a structure? At what time or number of cycles of loading will the crack propagate catastrophically? What size crack can be allowed to exist in component and still operate it safely? This last question may surprise you. Perhaps you would say that any crack, any flaw, is not allowable in the jet aircraft that carries your family across the ocean. Unfortunately such an aircraft does not exist. We must face reality square-on, recognize that flaws exist and to the very best of our ability, design our structures, monitoring protocols and maintenance procedures to ensure a low probability of failure by fracture. Doing so will save lives. Ignoring fracture could, in addition to the loss of life, bring down an entire corporation or industry and the livelihoods of thousands.

Fracture can and is being approached from many scales, [9]. For example at the atomic level, fracture can be viewed as the separation of atomic planes. At the scale of the microstructure of the material, the grains in a polycrystalline material, or the fibers in a composite, the fracture of the material around these features can be studied to determine the physical nature of failure. From the engineering point of view, the material is treated as a continuum and through the analysis of stress, strain and energy we seek to predict and control fracture. The continuum approach is the focus of this book.

Consider the example shown in Fig. 1.2. Here a sheet with initial crack length a is loaded with tensile stress σ_a. Near the crack tip the stress is elevated above the average stress of σ_a. Due to this high stress the material near the crack tip will undergo large strains and will eventually fail, allowing the crack to propagate ahead. If the material were to behave linearly elastically right up to the point of fracture then (as we will show in the next chapter) the stress ahead of the crack will be

$$\sigma_{22} = \frac{K_I}{\sqrt{2\pi r}}, \tag{1.1}$$

where r is the distance from the crack tip and K_I is related to the applied stress by $K_I = 1.12\sigma_a\sqrt{\pi a}$. The material will yield or otherwise inelastically and non-linearly deform to eliminate the predicted infinite stress, thus very near the crack tip Eq. (1.1) is not an accurate description of the stress field. However, if r_p, the size of the zone near the crack tip in which inelastic deformation occurs is small relative to a, the stress outside of this "yielding zone" will be well approximated by Eq. (1.1). This is the so called "small scale yielding" (SSY) assumption in fracture [10].

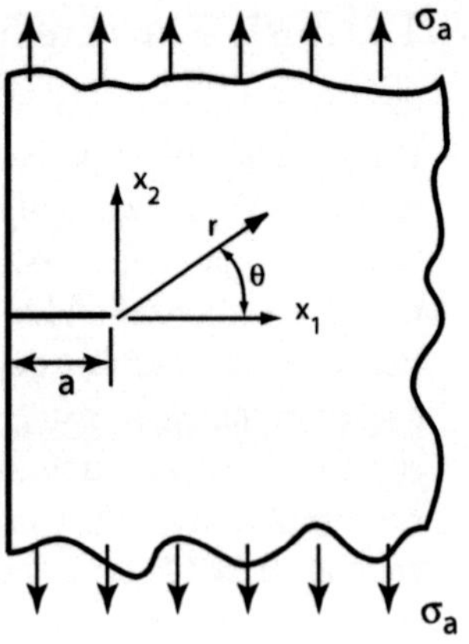

Fig. 1.2 Edge crack in a plate in tension. Mode I stress intensity factor, $K_I = 1.12\sigma_a\sqrt{\pi a}$

1.2.1 Small Scale Yielding Model

In the small scale yielding model the stresses in an annulus $r > r_p$ and $r \ll a$ are well approximated by $\sigma = \frac{K_I}{\sqrt{2\pi r}}\mathbf{f}(\theta)$ given with respect to polar coordinates, where $\mathbf{f}$ is a universal function of θ. All of the loading and geometry of loading are reflected in the single quantity K_I, known as the "stress intensity factor". That the distribution of stress around the crack tip has a universal spatial distribution with magnitude given by K_I is the so called "autonomy" principle. This allows fracture test results obtained from a 0.2 m laboratory test specimen to be applied to a 10 m large structure.

The size of the inelastic zone at the crack tip ("plastic zone", or "process zone") will be shown to scale as $r_p \sim K_I^2/\sigma_0^2$, where σ_0 is the yield strength of the material (the tensile stress at which inelastic deformation begins to occur).

Rice [10] (p. 217) describes the SSY yielding assumption and its role in fracture mechanics as:

> The utility of elastic stress analysis lies in the similarity of near crack tip stress distributions for all configurations. Presuming deviations from linearity to occur only over a region that is small compared to geometrical dimensions (small scale yielding), the elastic stress-intensity factor controls the local deformation field. This is in the sense that two bodies with cracks of different size and with different manners of load application, but which are otherwise identical, will have identical near crack tip deformation fields if the stress intensity factors are equal. Thus, the stress intensity factor uniquely characterizes the load sensed at the crack tip in situations of small scale yielding, and criteria governing crack extension for a given local load rate, temperature, environments, sheet thickness (where plane stress fracture modes are possible) and history of prior deformation may be expressed in terms of stress intensity factors.

1.2.2 Fracture Criteria

In SSY all crack tip deformation and failure is driven solely by K_I. A criterion for crack growth can be derived from this observation. The material has a characteristic resistance to fracture known as the "fracture toughness", K_{IC}. When the applied loading is such that

$$K_I \geq K_{IC}$$

then the crack will grow.

An alternate criterion for fracture is based on G, the "energy release rate", or energy dissipated per unit area of new fracture surface. As the crack grows in a component, work done on the component by the externally applied forces and strain energy stored in the part prior to fracture provide energy to the crack. The physical mechanisms of energy dissipation due to fracture include plastic deformation ahead of the crack in metals, microcracking in ceramics, fiber pull out and other frictional processes in composite materials, and surface energy in all materials. The surface energy component, is generally small relative to the other components, except in glassy materials. In the energy approach the criterion for fracture can be given as

$$G \geq G_C,$$

where G is the available energy release rate and G_C is the toughness of the materials, or energy per area required to propagate a crack.

It will be shown that in SSY the energy release rate, G is related to the stress intensity factor, K_I, by $G = \frac{K_I^2}{E'}$, where $E' = E$ for plane stress and $E' = E/(1-\nu^2)$ for plane strain and E is the Young's modulus of the material and ν the Poisson's ratio. Thus in SSY the stress intensity factor and energy release rate criteria are the same. This is not so, however when SSY is violated, which is generally the case for tearing fracture of ductile metals.

When the loading is applied cyclically and with $K_I < K_{IC}$ the material ahead of the crack will undergo fatigue deformation and eventually failure. It has been found that the crack will grow a small amount on each cycle of loading. The rate of crack growth typically scales as ΔK_I^n where ΔK is the difference between the maximum and minimum stress intensity factors due to the cyclic loads, and n is an exponent that must be experimentally determined. Typically $2 \leq n \leq 4$. Other situations in which a crack will grow slowly include stress corrosion cracking where under a constant $K_I < K_{IC}$ the crack can slowly advance as bonds are broken at the crack tip due to the interaction of stress with the corrosive agents. For example, you may have observed a crack slowly growing in an automobile windshield; water is known to catalyze fracture in glass.

1.3 Fracture Unit Conversions

1.0 ksi$\sqrt{\text{in}}$ = 1.099 MPa$\sqrt{\text{m}}$.

1.4 Exercises

1. Consider an aluminum plate loaded in tension. Suppose that the fracture toughness of this alloy is $K_{IC} \approx 60$ MPa$\sqrt{\text{m}}$ and the yield stress is $\sigma_y = 400$ MPa.

(a) If a tensile stress of $\sigma_a = 200$ MPa is applied what is the critical crack length, i.e. at what value of a is $K_I = K_{IC}$? At this critical crack length, estimate the size of the crack tip plastic zone using the relation $r_p = \frac{1}{\pi}\frac{K_I^2}{\sigma_y^2}$. Are the SSY conditions satisfied in this case?

2. Glass is a strong but very brittle material. Typically $K_{IC} \approx 1$ MPa$\sqrt{\text{m}}$ for glass. If the plate described above was made of glass and loaded in tension with $\sigma_a = 200$ MPa, what would the critical crack length be?

References

1. A.S. Svensson, Comet 1 world's first jetliner. http://w1.901.telia.com/u90113819/archives/comet.htm, last accessed 30 March, 2004 (2000)
2. W. Duncan, Engineering **179**, 196 (1955)
3. T. Bishop, *Metal Progress* (1955), pp. 79–85
4. E. Malnic, R. Meyer, Los Angeles Times, sec. 1(col. 6), p. 1 (1988)
5. D. Chandler, Boston Globe (1989), p. 1
6. S.A. Mahin, Lessons from steel buildings damaged during the Northridge earthquake. Tech. rep., National Information Service for Earthquake Engineering, University of California, Berkeley, http://nisee.berkeley.edu/northridge/mahin.html, last accessed 30 March 2004 (1997)
7. V. Esslinger, R. Kieselbach, R. Koller, B. Weisse, Eng. Fail. Anal. **11**, 515 (2004)
8. A. Butler, Aviat. Week Space Technol. **168**(2), 00052175 (2008)
9. T.J. Chang, J.W. Rudnicki (eds.), *Multiscale Deformation and Fracture in Materials and Structures the James R. Rice 60th Anniversary Volume*. Solid Mechanics and Its Applications, vol. 84 (Kluwer Academic, Dordrecht, 2000)
10. J.R. Rice, in *Fracture an Advanced Treatise*, vol. 2, ed. by H. Liebowitz (Academic Press, New York, 1968), pp. 191–311, chap. 3

Chapter 2
Linear Elastic Stress Analysis of 2D Cracks

Abstract To begin to understand fracture of materials, one must first know the stress and deformation fields near the tips of cracks. Thus the first topic in fracture mechanics is the linear elastic analysis of crack tip fields. The solutions derived here will be seen to violate the assumptions upon which linear elasticity theory is grounded. Nonetheless by invoking common sense principles, the theory of linear elastic fracture mechanics (LEFM) will be shown to provide the groundwork for many practical applications of fracture.

2.1 Notation

Unless otherwise stated all elastic analysis will be for static problems in linear elastic, isotropic, homogeneous materials in which no body forces act.

A *two dimensional* domain will be assumed to lie in the (x_1, x_2) plane and will be referred to as $\mathscr{A}$, with boundary curve $\mathscr{C}$ or Γ and outward unit normal vector $\mathbf{n}$. In a Cartesian coordinate system with basis vectors $\{\mathbf{e}_1, \mathbf{e}_2\}$, $\mathbf{n} = n_1\mathbf{e}_1 + n_2\mathbf{e}_2$, or $\mathbf{n} = n_\alpha \mathbf{e}_\alpha$ using the summation convention and the convention that Greek indices span 1, 2. An area integral will be denoted by $\int_{\mathscr{A}}(\cdot)dA$. A line integral is denoted by $\int_{\mathscr{C} \text{ or } \Gamma}(\cdot)d\Gamma$. New fracture surface area is referred to as $B \cdot da$, where B is the thickness of the 3D body that is idealized as 2D.

A three-dimensional domain will referred to as $\mathscr{V}$ with surface $\mathscr{S}$ and outward unit normal $\mathbf{n}$. The portion of the boundary over which tractions are prescribed is $\mathscr{S}_t$. The portion over which displacements are prescribed is $\mathscr{S}_u$. $\mathscr{S} = \mathscr{S}_t \bigcup \mathscr{S}_u$. In a Cartesian coordinate system with basis vectors $\{\mathbf{e}_1, \mathbf{e}_2, \mathbf{e}_3\}$, $\mathbf{n} = n_i\mathbf{e}_i$ where Latin indices span 1, 2, 3. A volume integral is denoted by $\int_{\mathscr{V}}(\cdot)dV$. A surface integral is denoted by $\int_{\mathscr{S}}(\cdot)dS$. New fracture surface area is referred to as ds or $\Delta\mathscr{S}$.

The stress tensor will be referred to as σ with components σ_{ij}. Strain is γ with components γ_{ij}. Traction $\mathbf{t} = \sigma\mathbf{n}$, or $t_i = \sigma_{ij}n_j$.

2.2 Introduction

Although real-world fracture problems involve crack surfaces that are curved and involve stress fields that are three dimensional, the only simple analyses that can

A.T. Zehnder, *Fracture Mechanics*,
Lecture Notes in Applied and Computational Mechanics 62,
DOI 10.1007/978-94-007-2595-9_2, © Springer Science+Business Media B.V. 2012

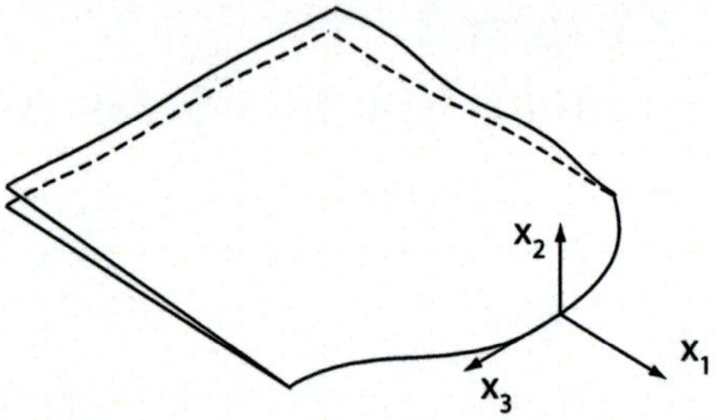

Fig. 2.1 Crack front, or line, for an arbitrarily shaped crack surface in a *solid*. At any point along the crack line a local coordinate system may be defined as shown

be performed are for two-dimensional idealizations. Solutions to these idealizations provide the basic structure of the crack tip fields.

Consider the arbitrary fracture surface shown in Fig. 2.1. At any point on the crack front a local coordinate system can be drawn with the x_3 axis tangential to the crack front, the x_2 axis orthogonal to the crack surface and x_1 orthogonal to the crack front. A polar coordinate system (r, θ) can be formed in the (x_1, x_2) plane. An observer who moves toward the crack tip along a path such that x_3 is constant will eventually be so close to the crack line front that the crack front appears to be straight and the crack surface flat. In such a case the three dimensional fracture problem at this point reduces to a two-dimensional one. The effects of the external loading and of the geometry of the problem are felt only through the magnitude and directions of the stress fields at the crack tip.

2.3 Modes of Fracture

At the crack tip the stress field can be broken up into three components, called Mode I, Mode II and Mode III, as sketched in Fig. 2.2, Mode I causes the crack to open orthogonal to the local fracture surface and results in tension or compressive stresses on surfaces that lie on the line $\theta = 0$ and that have normal vector $\mathbf{n} = \mathbf{e}_2$. Mode II causes the crack surfaces to slide relative to each other in the x_1 direction and results in shear stresses in the x_2 direction ahead of the crack. Mode-III causes the crack surface to slide relative to each other in the x_3 direction and results in shear stresses in the x_3 direction ahead of the crack.

With the idealization discussed above the solution of the crack tip fields can be broken down into three problems. Modes I and II are found by the solution of either a plane stress or plane strain problem and Mode III by the solution of an anti-plane shear problem.

2.4 Mode III Field

In many solid mechanics problems the anti-plane shear problem is the simplest to solve. This is also the case for fracture mechanics, thus we begin with this problem.

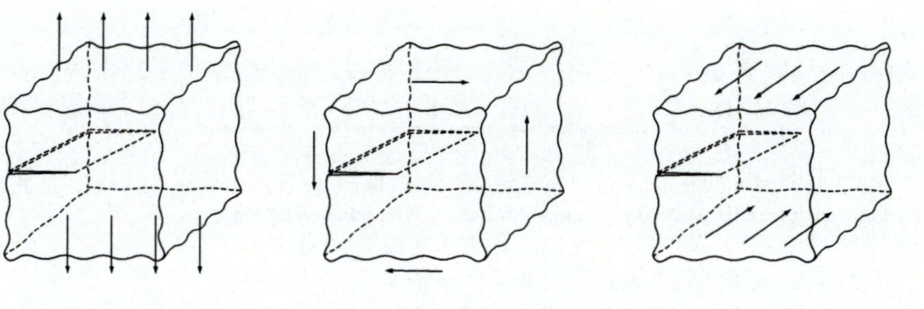

Fig. 2.2 Modes of fracture. Think of this as representing the state of stress for a cube of material surrounding part of a crack tip. The actual crack may have a mix of Mode-I,II,III loadings and this mix may vary along the crack front. The tractions on the front and back faces of Mode-III cube are not shown

Anti-plane shear is an idealization in which the displacement field is given by $\mathbf{u} = w(x_1, x_2)\mathbf{e}_3$. With this displacement field, the stress-strain relations are

$$\sigma_{3\alpha} = \mu w_{,\alpha}\,. \tag{2.1}$$

The field equations of linear elasticity reduce to

$$\nabla^2 w = 0 \tag{2.2}$$

on $\mathscr{A}$, with either traction boundary conditions,

$$\mu \nabla w \cdot \mathbf{n} = \sigma_{3\alpha} n_\alpha = \mu w_{,\alpha} n_\alpha = t_3^*(x_1, x_2) \tag{2.3}$$

or displacement boundary conditions,

$$w(x_1, w_2) = w^*(x_1, x_2) \tag{2.4}$$

on $\mathscr{C}$.

The anti-plane shear crack problem can be solved in two ways. In the first approach only the asymptotic fields near the crack tip are found. In the second, the entire stress field is found. Both solutions are given below.

2.4.1 Asymptotic Mode III Field

The geometry of the asymptotic problem is sketched in Fig. 2.3. An infinitely sharp, semi-infinite crack in an infinite body is assumed to lie along the x_1 axis. The crack surfaces are traction free.

This problem is best solved using polar coordinates, (r, θ). The field equation in polar coordinates is

$$\nabla^2 w = w_{,rr} + \frac{1}{r} w_{,r} + \frac{1}{r^2} w_{,\theta\theta} = 0, \tag{2.5}$$

Fig. 2.3 Semi-infinite crack in an infinite body. For clarity the crack is depicted with a small, but finite opening angle, actual problem is for a crack with no opening angle

and the traction free boundary conditions become

$$w_{,\theta}\,(r, \theta = \pm\pi) = 0. \tag{2.6}$$

Try to form a separable solution, $w(r,\theta) = R(r)T(\theta)$. Substituting into Eq. (2.5) and separating the r and θ dependent parts,

$$r^2\frac{R''}{R} + r\frac{R'}{R} = -\frac{T''}{T} = \left\{ \begin{array}{c} \lambda^2 \\ 0 \\ -\lambda^2 \end{array} \right\}, \tag{2.7}$$

where λ is a scalar. If the RHS of Eq. (2.7) is $-\lambda^2$ or 0 a trivial solution is obtained. Thus the only relevant case is when the RHS $= \lambda^2$. In this case the following two differential equations are obtained

$$T'' + \lambda^2 T = 0, \tag{2.8}$$

$$r^2 R'' + rR' - \lambda^2 R = 0. \tag{2.9}$$

The first has the solution

$$T(\theta) = A\cos\lambda\theta + B\sin\lambda\theta. \tag{2.10}$$

The second has the solution

$$R(r) = r^{\pm\lambda}. \tag{2.11}$$

The boundary conditions, $w_{,\theta}\,(r, \theta = \pm\pi) = 0$ become $R(r)T'(\pm\pi) = 0$. This leads to the pair of equations

$$\lambda(-A\sin\lambda\pi + B\cos\lambda\pi) = 0, \tag{2.12}$$

$$\lambda(A\sin\lambda\pi + B\cos\lambda\pi) = 0. \tag{2.13}$$

Adding and subtracting these equations leads to two sets of solutions

$$B\lambda\cos\lambda\pi = 0, \quad \Rightarrow \quad \lambda = 0, \lambda = \pm\, 1/2, \pm 3/2, \ldots, \tag{2.14}$$

$$A\lambda\sin\lambda\pi = 0, \quad \Rightarrow \quad \lambda = 0, \lambda = \pm 1, \pm 2, \ldots. \tag{2.15}$$

Thus the solution can be written as a series of terms. If $\lambda = 0$, then set $A = A_0$. Since $\lambda = 0$ corresponds to rigid body motion, set $B = 0$ when $\lambda = 0$ since it just adds to the A_0 term. If $\lambda = \pm 1/2, \pm 3/2, \ldots$, then from Eq. (2.12) $A = 0$. If $\lambda = \pm 1, \pm 2$, then $B = 0$.

Assembling the terms yields

$$w(r,\theta) = \sum_{n=-\infty}^{n=+\infty} A_n r^n \cos n\theta + B_n r^{n+\frac{1}{2}} \sin(n+1/2)\theta. \quad (2.16)$$

Noting that the stress field in polar coordinates is given by

$$\sigma_{3r} = \mu \frac{\partial w}{\partial r}, \qquad \sigma_{3\theta} = \frac{\mu}{r}\frac{\partial w}{\partial \theta}, \quad (2.17)$$

Eq. (2.12) predicts that the stress field is singular, i.e. the stress becomes infinitely large as $r \to 0$. Naturally this will also mean that the strain becomes infinite at the crack tip thus violating the small strain, linear theory of elasticity upon which the result is based.

Various arguments are traditionally used to restrict the terms in Eq. (2.12) to $n \geq 0$ resulting in a maximum stress singularity of $\sigma \sim r^{-1/2}$.

One argument is that the strain energy in a finite region must be bounded. In anti-plane shear the strain energy density is $W = \frac{\mu}{2}(w,_1^2 + w,_2^2)$. If $w \sim r^\lambda$, then $W \sim r^{2\lambda-2}$. The strain energy in a circular region of radius R surrounding the crack tip is

$$\int_{\lim r\to 0}^{R} \int_{-\pi}^{\pi} W r d\theta dr \sim R^{2\lambda} - \lim_{r\to 0} r^{2\lambda}.$$

Thus λ is restricted to $\lambda \geq 0$ for finite energy. From Eq. (2.12), $2\lambda = n$, or $n + 1/2$, thus if $\lambda \geq 0$ then we must restrict the series solution to $n \geq 0$.

A second argument is that the displacement must be bounded, which as with energy argument restricts the series to $n \geq 0$.

However, both of the above arguments assume the impossible, that the theory of linear elasticity is valid all the way to the crack tip despite the singular stresses. Even with the restriction that $n \geq 0$ the stress field is singular, thus since no material can sustain infinite stresses, there must exist a region surrounding the crack tip where the material yields or otherwise deforms nonlinearly in a way that relieves the stress singularity. If we don't claim that Eq. (2.12) must apply all the way to the crack tip, then outside of the crack tip nonlinear zone the energy and displacement will be finite for any order of singularity thus admitting terms with $n < 0$. Treating the crack tip nonlinear zone as a hole (an extreme model for material yielding in which the material's strength has dropped to zero) of radius ρ, Hui and Ruina [1] show that at any fixed, non-zero distance from the crack tip, the coefficients of terms with stresses more singular than $r^{-1/2}$ go to zero as $\rho/a \to 0$ where a is the crack length or other characteristic in-plane dimension such as the width of a test specimen or structural component. This result is in agreement with the restrictions placed on the crack tip fields by the energy and displacement arguments, thus in what follows the stress field is restricted to be no more singular than $\sigma \sim r^{-1/2}$. But note that in real-world problems in which the crack tip nonlinear zone is finite and $\rho/a \neq 0$, the stress field outside the nonlinear zone will have terms more singular than $r^{-1/2}$. Further details of this calculation are given in Sect. 7.3 as a prototype model for the effects of crack tip plasticity on the stress fields.

Based on the above arguments, and neglecting crack tip nonlinearities, all terms in the displacement series solution with negative powers of r are eliminated, leaving as the first four terms:

$$w(r,\theta) = A_0 + B_0 r^{1/2}\sin\frac{\theta}{2} + A_1 r\cos\theta + B_2 r^{3/2}\sin\frac{3\theta}{2} + \cdots. \tag{2.18}$$

Since the problem is a traction boundary value problem, the solution contains a rigid body motion term, A_0.

The stress field in polar coordinates is calculated by substituting Eq. (2.18) into Eq. (2.17), yielding

$$\sigma_{3r} = B_0\mu\frac{1}{2}r^{-1/2}\sin\frac{\theta}{2} + A_1\mu\cos\theta + B_2\mu\frac{3}{2}r^{1/2}\sin\frac{3\theta}{2} + \cdots, \tag{2.19}$$

$$\sigma_{3\theta} = B_0\mu\frac{1}{2}r^{-1/2}\cos\frac{\theta}{2} - A_1\mu\sin\theta + B_2\mu\frac{3}{2}r^{1/2}\cos\frac{3\theta}{2} + \cdots. \tag{2.20}$$

Note that the stress field has a characteristic $r^{-1/2}$ singularity. It will be shown that this singularity occurs for the Mode I and Mode II problems as well.

As $r \to 0$ the $r^{-1/2}$ term becomes much larger than the other terms in the series and the crack tip stress field is determined completely by B_0, the amplitude of the singular term. By convention the amplitude of the crack tip singularity is called the Mode III stress intensity factor, K_{III}, and is defined as

$$K_{III} \equiv \lim_{r\to 0}\sigma_{3\theta}(r,0)\sqrt{2\pi r}. \tag{2.21}$$

Substituting Eq. (2.20) into the above, B_0 can be written as $B_0 = \sqrt{\frac{2}{\pi}}\frac{K_{III}}{\mu}$. Using the language of stress intensity factors, the first three terms of the series solution for the displacement and stress fields can be written as

$$w(r,\theta) = A_0 + \sqrt{\frac{2}{\pi}}\frac{K_{III}}{\mu}r^{1/2}\sin\frac{\theta}{2} + A_1 r\cos\theta + B_2 r^{3/2}\sin\frac{3\theta}{2} + \cdots \tag{2.22}$$

and

$$\begin{pmatrix}\sigma_{3r}\\ \sigma_{3\theta}\end{pmatrix} = \frac{K_{III}}{\sqrt{2\pi r}}\begin{pmatrix}\sin\frac{\theta}{2}\\ \cos\frac{\theta}{2}\end{pmatrix} + A_1\mu\begin{pmatrix}\cos\theta\\ -\sin\theta\end{pmatrix} + \frac{3B_2\mu r^{1/2}}{2}\begin{pmatrix}\sin\frac{3\theta}{2}\\ \cos\frac{3\theta}{2}\end{pmatrix}. \tag{2.23}$$

The stress intensity factor, K_{III} is not determined from this analysis. In general K_{III} will depend linearly on the applied loads and will also depend on the specific geometry of the cracked body and on the distribution of loads. There are a number approaches to calculating the stress intensity factor, many of which will be discussed later in this book.

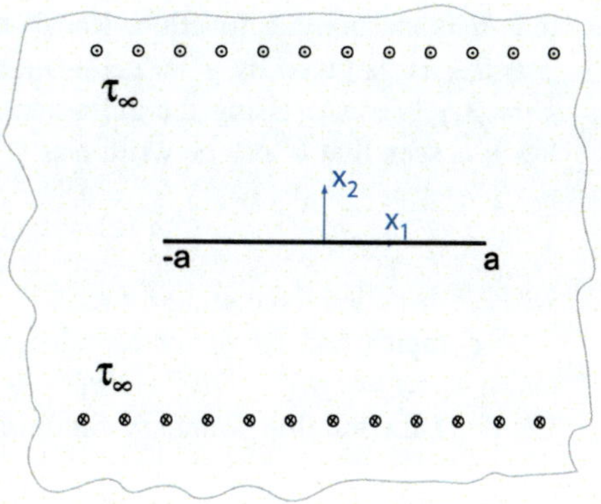

Fig. 2.4 Finite crack of length $2a$ in an infinite body under uniform anti-plane shear loading in the far field

2.4.2 Full Field for Finite Crack in an Infinite Body

A crack that is small compared to the plate dimension and whose shortest ligament from the crack to the outer plate boundary is much larger than the crack can be approximated as a finite crack in an infinite plate. If, in addition, the spatial variation of the stress field is not large, such a problem may be modeled as a crack of length $2a$ loaded by uniform shear stresses, $\sigma_{31} = 0, \sigma_{32} = \tau_\infty$, Fig. 2.4.

2.4.2.1 Complex Variables Formulation of Anti-Plane Shear

To simplify the notation the following definitions are made: $\tau_\alpha = \sigma_{3\alpha}, \gamma_\alpha = 2\gamma_{3\alpha}$. Let χ be a stress function such that

$$\tau_1 = -\frac{\partial \chi}{\partial x_2}, \quad \text{and} \quad \tau_2 = \frac{\partial \chi}{\partial x_1}. \tag{2.24}$$

From the strain-displacement relations $\gamma_\alpha = w,_\alpha$. Thus $\gamma_{1,2} = w,_{12}$ and $\gamma_{2,1} = w,_{21}$ from which the compatibility relation

$$\gamma_{1,2} = \gamma_{2,1} \tag{2.25}$$

is obtained. Using the stress strain relations, $\tau_\alpha = \mu\gamma_\alpha$, and the stress functions yields $\mu\gamma_{1,2} = -\chi,_{22}$ and $\mu\gamma_{2,1} = \chi,_{11}$. Substituting this into the compatibility equation yields $-\chi,_{22} = \chi,_{11}$ or

$$\nabla^2 \chi = 0. \tag{2.26}$$

Define a new, complex function using χ as the real part and w as the imaginary part,

$$h(z) = \chi + i\mu w \tag{2.27}$$

where $z = x_1 + ix_2$. It is easily verified that χ and w satisfy the Cauchy-Riemann equations. Furthermore both χ and w are harmonic, i.e. $\nabla^2 \chi = 0$ and $\nabla^2 w = 0$, thus h is an analytic function. Recall that the derivative of an analytic function, $f = u + iv$ is given by $f' = u_{,1} + iv_1 = v_{,2} - iu_{,2}$. Applying this rule to h yields $h' = \chi_{,1} + i\mu w_{,1}$. Using the definition of the stress function and the stress-strain law it is seen that h' can be written as

$$h'(z) = \tau_2(z) + i\tau_1(z) \equiv \tau \tag{2.28}$$

where τ is called the complex stress.

A complex normal vector can also be defined, $n \equiv n_1 + in_2$. The product of τ and n is $\tau n = \tau_2 n_1 - \tau_1 n_2 + i(\tau_1 n_1 + \tau_2 n_2)$. Thus, comparing this expression to Eq. (2.3), the traction boundary conditions can be written as

$$\mathrm{Im}[\tau(z)n(z)] = t^*(z) \tag{2.29}$$

on $\mathscr{C}$.

2.4.2.2 Solution to the Problem

The problem to be solved is outlined in Fig. 2.4. A finite crack of length $2a$ lies along the x_1 axis. Far away from the crack a uniform shear stress field is applied, $\tau_1 = 0$, $\tau_2 = \tau_\infty$, or in terms of the complex stress, $\tau = \tau_\infty + i0$. The crack surfaces are traction free, i.e. $\mathrm{Re}[\tau] = \tau_2 = 0$ on $-a \le x_1 \le a$, $x_2 = 0$.

This problem can be solved by analogy to the solution for fluid flow around a flat plate, [2]. In the fluid problem the flow velocity v is given by $v = \overline{F'(z)}$, where $F = A(z^2 - a^2)^{1/2}$. With the fluid velocity analogous to the stress, try a solution of the form

$$h(z) = A(z^2 - a^2)^{1/2}. \tag{2.30}$$

It is easily shown that for $z \ne \pm a$ this function is analytic, thus the governing pde for anti-plane shear will be satisfied. All that remains is to check if the boundary conditions are satisfied. With the above h, the complex stress is

$$\tau = h'(z) = \frac{Az}{(z^2 - a^2)^{1/2}}. \tag{2.31}$$

As $z \to \infty$ $\tau \to A$, thus to satisfy the far-field boundary condition $A = \tau_\infty$.

To check if the crack tip is traction free note that in reference to Fig. 2.5 $z - a = r_1 e^{i\theta_1}$ and $z + a = r_2 e^{i\theta_2}$. Thus $z^2 - a^2 = r_1 r_2 e^{i(\theta_1 + \theta_2)}$.

On the top crack surface, $x_2 = 0^+$, $-a \le x_1 \le a$, $\theta_1 = \pi$ and $\theta_2 = 0$, thus $z^2 - a^2 = r_1 r_2 e^{i(\pi+0)} = -r_1 r_2 = -a^2 + x_1^2$. Thus on this surface the complex stress is $\tau = \frac{\tau_\infty x_1}{\sqrt{-r_1 r_2}} = \frac{-i\tau_\infty x_1}{\sqrt{a^2 - x^2}}$. The traction free boundary condition on the top crack surface is $\mathrm{Im}[\tau n] = 0$ where $n = i$, thus the boundary condition can be written as $\mathrm{Re}[\tau] = 0$. Since the complex stress on the top fracture surface has only an imaginary part, the traction free boundary condition is shown to be satisfied.

On the bottom crack surface, $x_2 = 0^-$, $-a \le x_1 \le a$, $\theta_1 = \pi$ and $\theta_2 = 2\pi$, thus $z^2 - a^2 = r_1 r_2 e^{i(\pi+2\pi)} = -r_1 r_2 = -a^2 + x_1^2$ and again the stress has no real part, thus showing that the traction free boundary conditions will be satisfied.

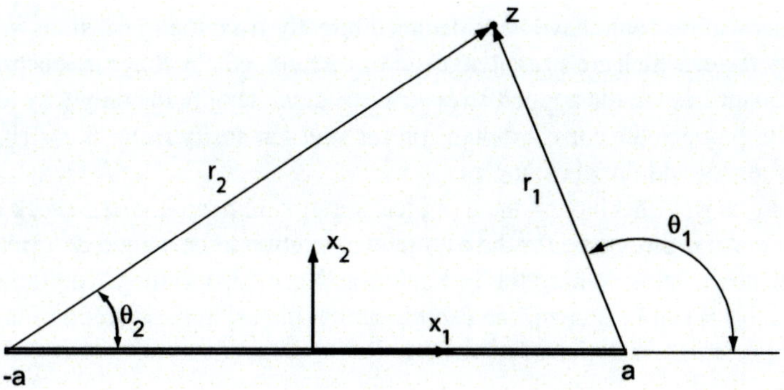

Fig. 2.5 Finite, antiplane-shear crack in an infinite body. θ_1 is discontinuous along $z = x_1, x_1 \geq a$. θ_2 is discontinuous along $z = x_1, x_1 \geq -a$

To summarize we have the following displacement and stress fields

$$w = \frac{1}{\mu}\,\mathrm{Im}[h] = \mathrm{Im}\,\frac{\tau_\infty}{\mu}\sqrt{(z^2 - a^2)}, \tag{2.32}$$

$$\tau = \tau_2 + i\tau_1 = \frac{\tau_\infty z}{\sqrt{z^2 - a^2}}. \tag{2.33}$$

Your intuition will tell you that near the crack tip this solution should give the same result as Eq. (2.23). To show that this is so, the stress field is analyzed near the right crack tip, $z \to a$. Note that $z^2 - a^2 = (z + a)(z - a)$. Setting $z \approx a$, $z^2 - a^2 \approx (z - a)(2a)$, hence near the right hand crack tip $\tau = \frac{\tau_\infty a}{\sqrt{z-a}\sqrt{2a}}$. Writing $z - a = r_1 e^{i\theta_1}$, and relabeling $\theta_1 = \theta$ the stress can be written as

$$\tau = \tau_2 + i\tau_1 = \frac{\tau_\infty\sqrt{a}}{\sqrt{2}\sqrt{r_1}}e^{-i\theta/2} = \frac{\tau_\infty\sqrt{a}}{\sqrt{2r_1}}\big(\cos\theta/2 - i\sin\theta/2\big) \quad \text{as } r_1 \to 0. \tag{2.34}$$

Comparing the above to Eq. (2.23) it is verified that near the crack tip the two stress fields are the same.

Note that unlike the asymptotic problem, the stress field in this problem is completely determined and the stress intensity factor can be determined. Recall the definition of the Mode-III stress intensity factor, Eq. (2.21) $K_{III} \equiv \lim_{r\to 0}\sigma_{3\theta}(r, 0)\sqrt{2\pi r}$. Noting that τ_2 is simply a shorthand notation for σ_{32}, and that r in the asymptotic problem is the same as r_1 in the finite crack problem, from Eq. (2.34)

$$K_{III} = \frac{\tau_\infty\sqrt{a}}{\sqrt{2r_1}}\sqrt{2\pi r_1} = \tau_\infty\sqrt{\pi a}. \tag{2.35}$$

Thus it is seen that the stress intensity factor scales as the applied load (τ_∞) and the square root of the crack length (a). As other problems are discussed it will be seen that such scaling arises again and again.

This scaling could have been deduced directly from the dimensions of stress intensity factor which are stress·length$^{1/2}$ or force/length$^{3/2}$. Since in this problem the only quantities are the applied stress and the crack length, the only way to combine them to produce the correct dimension for stress intensity factor is $\tau_\infty a^{1/2}$. See the exercises for additional examples.

Note as well that having the complete solution in hand one can check how close to the crack must one be for the asymptotic solution to be a good description of the actual stress fields. Taking the full solution, Eq. (2.33) to the asymptotic solution, Eq. (2.34) it can be shown, see exercises, that the asymptotic solution is valid in a region near the crack tips of $r \preceq a/10$.

2.5 Mode I and Mode II Fields

As with the Mode III field, the Mode I and Mode II problems can be solved either by asymptotic analysis or through the solution to a specific boundary value problem such as a finite crack in an infinite plate. However, as in the analysis above for the Mode III crack, the near crack tip stress fields are the same in each case. Thus the approach of calculating only the asymptotic stress fields will be taken here, following the analysis of Williams [3].

The Mode-I and Mode-II problems are sketched in Fig. 2.2. The coordinate system and geometry are the same as the Mode-III asymptotic problem, Fig. 2.3. Plane stress and plane strain are assumed.

2.5.1 Review of Plane Stress and Plane Strain Field Equations

2.5.1.1 Plane Strain

The plane strain assumption is that $u_3 = 0$ and $u_\alpha = u_\alpha(x_1, x_2)$. This assumption is appropriate for plane problems in which the loading is all in the x_1, x_2 plane and for bodies in which the thickness (x_3 direction) is much greater than the in-plane (x_1, x_2) dimensions. The reader can refer to an textbook on linear elasticity theory for the derivations of the following results:

$$
\begin{aligned}
\gamma_{\alpha\beta} &= \frac{1}{2}(u_{\alpha,\beta} + u_{\beta,\alpha}),\\
\gamma_{\alpha\beta} &= \frac{1+\nu}{E}(\sigma_{\alpha\beta} - \nu\sigma_{\gamma\gamma}\delta_{\alpha\beta}),\\
\sigma_{\alpha\beta,\beta} &= 0,\\
\sigma_{33} &= \nu\sigma_{\gamma\gamma}.
\end{aligned}
$$

2.5.1.2 Plane Stress

The plane stress assumption is that $\sigma_{33}=0$ and that $u_\alpha = u_\alpha(x_1, x_2)$. This assumption is appropriate for plane problems in bodies that are thin relative to their in-plane dimensions. For example, the fields for crack in a plate of thin sheet metal loaded in tension could be well approximated by a plane stress solution. The strain-displacement and equilibrium equations are the same as for plane strain. The stress-strain law can be written as

$$\gamma_{33} = -\frac{\nu}{E}\sigma_{\gamma\gamma} = -\frac{\nu}{1-\nu}\gamma_{\gamma\gamma},$$

$$\gamma_{\alpha\beta} = \frac{1+\nu}{E}\left(\sigma_{\alpha\beta} - \frac{\nu}{1+\nu}\sigma_{\gamma\gamma}\delta_{\alpha\beta}\right).$$

2.5.1.3 Stress Function

To solve for the stress field one approach is to define and then solve for the stress function, Φ. In Cartesian coordinates the stresses are related to $\Phi(x_1, x_2)$ by

$$\sigma_{11} = \Phi_{,22}, \tag{2.36}$$

$$\sigma_{22} = \Phi_{,11}, \tag{2.37}$$

$$\sigma_{12} = -\Phi_{,12}. \tag{2.38}$$

In polar coordinates the stress is related to $\Phi(r, \theta)$ by

$$\sigma_{\theta\theta} = \Phi_{,rr}, \tag{2.39}$$

$$\sigma_{rr} = \frac{1}{r}\Phi_{,r} + \frac{1}{r^2}\Phi_{,\theta\theta}, \tag{2.40}$$

$$\sigma_{r\theta} = -\left(\frac{1}{r}\Phi_{,\theta}\right)_{,r}. \tag{2.41}$$

It is readily shown that stresses derived from such a stress function satisfy the equilibrium equations. Requiring the stresses to satisfy compatibility requires that Φ satisfies the biharmonic equation

$$\nabla^4\Phi = 0. \tag{2.42}$$

In polar coordinates this can be written as $\nabla^4\Phi = \nabla^2(\nabla^2\Phi)$, $\nabla^2\Phi = \Phi_{,rr} + \frac{1}{r}\Phi_{,r} + \frac{1}{r^2}\Phi_{,\theta\theta}$.

2.5.2 Asymptotic Mode I Field

2.5.2.1 Stress Field

The asymptotic crack problem is the same as that shown in Fig. 2.3. The traction free boundary conditions, $\mathbf{t}=0$ on $\theta = \pm\pi$ require that $\sigma_{\theta\theta} = \sigma_{r\theta} = 0$ on $\theta = \pm\pi$. In terms of the stress function the boundary conditions are $\Phi,_{rr} = 0$ and $(\frac{1}{r}\Phi,_{\theta}),_r = 0$ on $\theta = \pm\pi$.

Following Williams's approach [3] consider a solution of the form

$$\begin{aligned}\Phi(r,\theta) &= r^{\lambda+2}\big[A\cos\lambda\theta + B\cos(\lambda+2)\theta\big] \\ &\quad + r^{\lambda+2}\big[C\sin\lambda\theta + D\sin(\lambda+2)\theta\big].\end{aligned} \tag{2.43}$$

Note that one could start from a more basic approach. For example the general solution to the biharmonic equation in polar coordinates, found in 1899 by Michell and given in Timoshenko and Goodier [4] could be used as a starting point. Only certain terms of this result, corresponding to those used by Williams, will be needed to satisfy the boundary conditions of the crack problem.

It will be noted that the first two terms of Eq. (2.43) are symmetric with respect to the crack line and that the second two are anti-symmetric with respect to the crack. It will be shown that these correspond to the solutions of the Mode-I and Mode-II problems respectively. Let us consider for now, only the Mode-I solution. The boundary condition $\Phi, rr = 0$ on $\pm\pi$ (normal component of traction) yields

$$\Phi,_{rr}|_{\pi} = (\lambda+2)(\lambda+1)r^{\lambda}\big[A\cos\lambda\pi + B\cos(\lambda+2)\pi\big] = 0.$$

Noting that $\cos(\lambda\pi + 2\pi) = \cos\lambda\pi$, the above requires (for a nontrivial solution) that

$$(\lambda+2)(\lambda+1)(A+B)\cos\lambda\pi = 0. \tag{2.44}$$

The requirement that the shear component of traction is zero, yields

$$\left(\frac{1}{r}\Phi,_{\theta}\right),_r\bigg|_{\pi} = (\lambda+1)r^{\lambda}\big[-A\lambda\sin\lambda\pi - B(\lambda+2)\sin(\lambda\pi+2\pi)\big] = 0.$$

This leads to

$$\sin\lambda\pi\big[A\lambda + B(\lambda+2)\big]. \tag{2.45}$$

If the stress function is $\Phi \sim r^{\lambda+2}$, then the stress will be $\sigma \sim r^{\lambda}$, and since stress and strain are proportional to the first derivatives of the displacement, the displacement fields will be $u \sim r^{\lambda+1}$, $\lambda \neq -1$ or $u \sim \ln r$, $\lambda = -1$. As in the anti-plane shear problem, a reasonable assumption is that the displacements at the crack tip will be finite. This will restrict the solution to $\lambda > -1$.

To satisfy Eqs. (2.44) and (2.45) requires that

$$\cos\lambda\pi = 0 \quad\Rightarrow\quad \lambda = -\frac{1}{2}, \frac{1}{2}, \frac{3}{2}, \ldots, \quad \text{and} \quad B = -\lambda A/(\lambda+2),$$

or

$$\sin\lambda\pi = 0 \quad\Rightarrow\quad \lambda = 0, 1, 2, \ldots, \quad \text{and} \quad B = -A.$$

Taking the first three terms of the solution, for $\lambda = -\frac{1}{2}$, $B_{-1/2} = \frac{1}{3}A_{-1/2}$, for $\lambda = 0$, $B_0 = -A_0$ and for $\lambda = 1/2$, $B_{1/2} = -\frac{1}{5}A_{1/2}$. Thus the stress function is

$$\Phi(r,\theta) = r^{3/2}A_{-1/2}\left[\cos\frac{\theta}{2} + \frac{1}{3}\cos\frac{3\theta}{2}\right] + r^2 A_0[1-\cos 2\theta] + r^{5/2}A_{1/2}\left[\cos\frac{\theta}{2} - \frac{1}{5}\cos\frac{5\theta}{2}\right] + \text{H.O.T.} \tag{2.46}$$

Taking the derivative $\Phi_{,rr}$, the "hoop stress", $\sigma_{\theta\theta}$ is

$$\sigma_{\theta\theta} = \frac{3}{4}A_{-1/2}r^{-1/2}\left[\cos\frac{\theta}{2} + \frac{1}{3}\cos\frac{3\theta}{2}\right] + 2A_0\left[1-\cos 2\theta\right] + \frac{1}{4}A_{1/2}r^{1/2}\left[15\cos\frac{\theta}{2} - 3\cos\frac{5\theta}{2}\right] + \text{H.O.T.}$$

As in the anti-plane shear problem, the crack tip stress field is infinite with a $1/\sqrt{r}$ singularity. The strength of this singularity is given by the "Mode-I" stress intensity factor, K_I. By definition,

$$K_I \equiv \lim_{r\to 0}\sigma_{\theta\theta}|_{\theta=0}\sqrt{2\pi r} = A_{-1/2}\sqrt{2\pi}. \tag{2.47}$$

Replacing $A_{-1/2}$ by $K_I/\sqrt{2\pi}$ the stress function can be written as

$$\Phi(r,\theta) = \frac{K_I}{\sqrt{2\pi}}r^{3/2}\left[\cos\frac{\theta}{2} + \frac{1}{3}\cos\frac{3\theta}{2}\right] + 2A_0 r^2[1-\cos 2\theta] + r^{5/2}A_{1/2}\left[\cos\frac{\theta}{2} - \frac{1}{5}\cos\frac{5\theta}{2}\right] + \text{H.O.T.} \tag{2.48}$$

Taking derivatives of the stress function, the stress field can be written as

$$\begin{pmatrix}\sigma_{rr}\\ \sigma_{\theta\theta}\\ \sigma_{r\theta}\end{pmatrix} = \frac{K_I}{\sqrt{2\pi r}}\frac{1}{4}\begin{pmatrix}-\cos\frac{3\theta}{2} + 5\cos\frac{\theta}{2}\\ \cos\frac{3\theta}{2} + 3\cos\frac{\theta}{2}\\ \sin\frac{\theta}{2} + \sin\frac{3\theta}{2}\end{pmatrix} + 4A_0\begin{pmatrix}\cos^2\theta\\ \sin^2\theta\\ -\sin\theta\cos\theta\end{pmatrix} + \frac{3A_{1/2}r^{1/2}}{4}\begin{pmatrix}3\cos\frac{\theta}{2} + \cos\frac{5\theta}{2}\\ 5\cos\frac{\theta}{2} - \cos\frac{5\theta}{2}\\ \sin\frac{\theta}{2} - \sin\frac{5\theta}{2}\end{pmatrix} + \text{H.O.T.} \tag{2.49}$$

2.5.2.2 Displacement Field

Finding the displacement field can be a more difficult problem than finding the stress field. One approach is to calculate the strains using the stress-strain laws, and then integrate the strain-displacement relations to determine the displacement fields. Williams used the approach of starting from the solution of Coker and Filon [5] in which it is shown that the displacement components in polar coordinates are related to the stress function by

$$2\mu u_r = -\Phi,_r + (1-\overline{\nu})r\Psi,_\theta\,, \tag{2.50}$$

$$2\mu u_\theta = -\frac{1}{r}\Phi,_\theta + (1-\overline{\nu})r^2\Psi,_r \tag{2.51}$$

where the displacement potential, Ψ is related to the stress function by

$$\nabla^2\Phi = (r\Psi,_\theta),_r\,, \tag{2.52}$$

μ is the shear modulus, and $\overline{\nu} = \nu$ for plane strain and $\overline{\nu} = \nu/(1+\nu)$ for plane stress.

As above, the (Mode-I) stress function is a power series in r. Assume that the displacement potential can also we written as a power series, thus we have

$$\Phi(r,\theta) = r^{\lambda+2}[A\cos\lambda\theta + B\cos(\lambda+2)\theta], \tag{2.53}$$

$$\Psi(r,\theta) = r^m[a_1\cos m\theta + a_2\sin m\theta]. \tag{2.54}$$

Evaluating the derivatives of Eq. (2.53) and substituting into Eq. (2.52) yields $a_1 = 0$, $a_2 = 4A/\lambda$ and $m = \lambda$. Thus the terms of the Mode-I displacement potential are

$$\Psi = r^\lambda \frac{4A}{\lambda}\sin\lambda\theta. \tag{2.55}$$

Taking only the first term of the series (corresponding to $\lambda = -1/2$),

$$\Phi = A_{-1/2}r^{3/2}\left[\cos\frac{\theta}{2} + \frac{1}{3}\cos\frac{3\theta}{2}\right], \tag{2.56}$$

$$\Psi = 8A_{-1/2}r^{-1/2}\sin\frac{\theta}{2}. \tag{2.57}$$

Substituting into Eq. (2.50) and replacing $A_{-1/2}$ by $K_I/\sqrt{2\pi}$ yields

$$\begin{pmatrix} u_r \\ u_\theta \end{pmatrix} = K_I\frac{(1+\nu)}{E}\sqrt{\frac{r}{2\pi}}\begin{pmatrix} (\frac{5}{2}-4\overline{\nu})\cos\frac{\theta}{3} - \frac{1}{2}\cos\frac{3\theta}{2} \\ -(\frac{7}{2}-4\overline{\nu})\sin\frac{\theta}{2} + \frac{1}{2}\sin\frac{3\theta}{2} \end{pmatrix}. \tag{2.58}$$

The shape of the crack under load is a parabola, as can be found by considering the opening displacement of the crack, $u_2(r,\pm\pi) = -u_\theta(r,\pm\pi)$:

$$u_2(r,\pm\pi) = -u_\theta(r,\pm\pi) = \pm\frac{4K_I}{E'}\sqrt{\frac{r}{2\pi}} \tag{2.59}$$

where $E' = E$ for plane stress and $E' = \frac{E}{1-\nu^2}$ for plane strain.

2.5.3 Asymptotic Mode II Field

The details of the Mode II solution will not be given as the steps are identical to those taken for the Mode I solution. The resulting stress and displacement fields are expressed in terms of the Mode-II stress intensity factor, K_{II}, defined as

$$K_{II} \equiv \lim_{r\to 0} \sigma_{r\theta}|_{\theta=0}\sqrt{2\pi r}. \tag{2.60}$$

The first term of the stress field is given by

$$\begin{pmatrix} \sigma_{rr} \\ \sigma_{\theta\theta} \\ \sigma_{r\theta} \end{pmatrix} = \frac{K_{II}}{\sqrt{2\pi r}}\frac{1}{4}\begin{pmatrix} -5\sin\frac{\theta}{2} + 3\sin\frac{3\theta}{2} \\ -3\sin\frac{\theta}{2} - 3\sin\frac{3\theta}{2} \\ \cos\frac{\theta}{2} + 3\cos\frac{3\theta}{2} \end{pmatrix}. \tag{2.61}$$

The displacement field is given by

$$\begin{pmatrix} u_r \\ u_\theta \end{pmatrix} = K_{II}\frac{(1+\nu)}{E}\sqrt{\frac{r}{2\pi}}\begin{pmatrix} (-\frac{5}{2} + 4\overline{\nu})\sin\frac{\theta}{2} + \frac{3}{2}\sin\frac{3\theta}{2} \\ -(\frac{7}{2} - 4\overline{\nu})\cos\frac{\theta}{2} + \frac{3}{2}\cos\frac{3\theta}{2} \end{pmatrix}. \tag{2.62}$$

2.6 Complex Variables Method for Mode I and Mode II Cracks

To determine the full stress field for a finite Mode-I or Mode-II crack we will need to use the method of complex variables. The solution we develop will allows us to find the stress and displacement fields as well as the stress intensity factors for any loading of a finite crack in an infinite plate. We consider a crack of length $2a$ lying along $x_2 = 0$, as shown in Fig. 2.5.

Following Hellan [6], the biharmonic equation (2.42) $\nabla^4 \Phi = 0$, is solved by

$$2\Phi = \mathrm{Re}\big[\bar{z}\phi(z) + \psi(z)\big], \tag{2.63}$$

where ϕ and ψ are analytic functions of $z = x_1 + ix_2$. The stresses are given by

$$\sigma_{11} = \mathrm{Re}\left[\phi' - \frac{1}{2}\bar{z}\phi'' - \frac{1}{2}\psi''\right], \tag{2.64}$$

$$\sigma_{22} = \mathrm{Re}\left[\phi' + \frac{1}{2}\bar{z}\phi'' + \frac{1}{2}\psi''\right], \tag{2.65}$$

$$\sigma_{12} = \frac{1}{2}\mathrm{Im}\big[\bar{z}\phi'' + \psi''\big]. \tag{2.66}$$

The displacements can be found from

$$4\mu u_1 = \mathrm{Re}\big[\kappa\phi - \bar{z}\phi' - \psi'\big], \tag{2.67}$$

$$4\mu u_2 = \mathrm{Im}\big[\kappa\phi + \bar{z}\phi' + \psi'\big], \tag{2.68}$$

where $\kappa = 3 - 4\nu$ for plane strain and $\kappa = (3-\nu)/(1+\nu)$ for plane stress.

The fracture problems can be broken up into Mode-I (symmetric) and Mode-II (anti-symmetric) problems. To simplify the calculations the results above can be specialized to the two cases using the Westergaard approach [7].

2.6.1 Westergaard Approach for Mode-I

For the Mode-I case, along $x_2 = 0$ $\sigma_{12} = 0$, which can be enforced by setting $\psi'' = -z\phi''$. In this case $\psi' = -z\phi' + \phi + \text{const.}$ and the stresses can be written as

$$\begin{aligned} \sigma_{11} &= \operatorname{Re}\phi' - x_2 \operatorname{Im}\phi'', \\ \sigma_{22} &= \operatorname{Re}\phi' + x_2 \operatorname{Im}\phi'', \\ \sigma_{12} &= -x_2 \operatorname{Re}\phi''. \end{aligned} \tag{2.69}$$

The displacements are

$$\begin{aligned} 2\mu u_1 &= \frac{\kappa - 1}{2} \operatorname{Re}\phi - x_2 \operatorname{Im}\phi', \\ 2\mu u_2 &= \frac{\kappa + 1}{2} \operatorname{Im}\phi - x_2 \operatorname{Re}\phi'. \end{aligned} \tag{2.70}$$

2.6.2 Westergaard Approach for Mode-II

For Mode-II, along $x_2 = 0$ $\sigma_{22} = 0$ which can be enforced by setting $\psi'' = -2\phi' - z\phi''$. In this case $\psi' = -\phi - z\phi' + \text{const.}$ The stresses are

$$\begin{aligned} \sigma_{11} &= 2\operatorname{Re}\phi' - x_2 \operatorname{Im}\phi'', \\ \sigma_{22} &= x_2 \operatorname{Im}\phi'', \\ \sigma_{12} &= -\operatorname{Im}\phi' - x_2 \operatorname{Re}\phi''. \end{aligned} \tag{2.71}$$

The displacements are

$$\begin{aligned} 2\mu u_1 &= \frac{\kappa + 1}{2} \operatorname{Re}\phi - x_2 \operatorname{Im}\phi', \\ 2\mu u_2 &= \frac{\kappa - 1}{2} \operatorname{Im}\phi - x_2 \operatorname{Re}\phi'. \end{aligned} \tag{2.72}$$

2.6.3 General Solution for Internal Crack with Applied Tractions

If the crack surfaces have traction loading $\mathbf{t} = p_1(x_1)\mathbf{e}_1 + p_2(x_1)\mathbf{e}_2$ on the top surface and equal but opposite tractions on the bottom surface, as shown in Fig. 2.6 Sedov [8] gives the following general solutions for ϕ'.

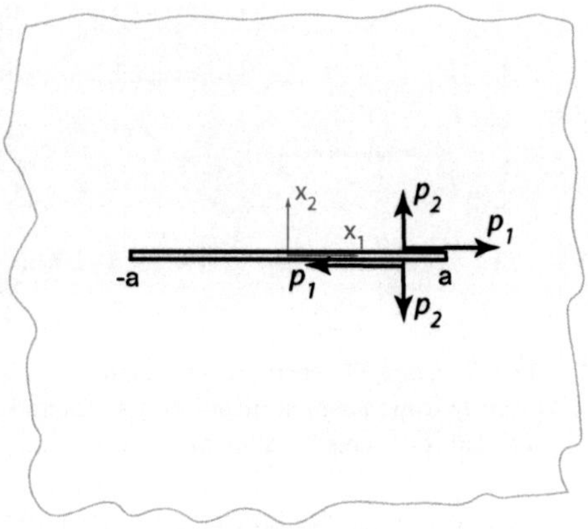

Fig. 2.6 Traction on crack face

For Mode-I,

$$\phi' = \frac{1}{\pi\sqrt{z^2 - a^2}} \int_{-a}^{a} p_2(t) \frac{\sqrt{a^2 - t^2}}{z - t} dt. \tag{2.73}$$

For Mode-II,

$$\phi' = \frac{-i}{\pi\sqrt{z^2 - a^2}} \int_{-a}^{a} p_1(t) \frac{\sqrt{a^2 - t^2}}{z - t} dt. \tag{2.74}$$

2.6.4 Full Stress Field for Mode-I Crack in an Infinite Plate

The stress and displacement fields for a finite crack subject to uniform tension loading, $\sigma_{22} = \sigma_\infty$, $\sigma_{11} = 0$, and $\sigma_{12} = 0$ can now be calculated using the above method. A superposition approach is taken as sketched in Fig. 2.7. If no crack were present, than along $x_2 = 0$ there would be a tensile stress of $\sigma_{22} = \sigma^\infty$. To make the crack traction free we apply a compressive stress to the crack faces, i.e. on the upper crack facc apply $p_2 = \sigma_\infty$. The solution to the problem is the superposition of the uniform stress $\sigma_{22} = \sigma_\infty$ with the stress due to the crack face loadings.

For the crack face loading part of the problem,

$$\phi' = \frac{\sigma_\infty}{\pi\sqrt{z^2 - a^2}} \int_{-a}^{a} \frac{\sqrt{a^2 - t^2}}{z - t} dt. \tag{2.75}$$

Evaluating this integral yields

$$\phi' = \frac{\sigma_\infty z}{\sqrt{z^2 - a^2}} - \sigma_\infty, \tag{2.76}$$

which can be integrated to yield

$$\phi = \sigma_\infty\sqrt{z^2 - a^2} - \sigma_\infty z + \text{const.} \tag{2.77}$$

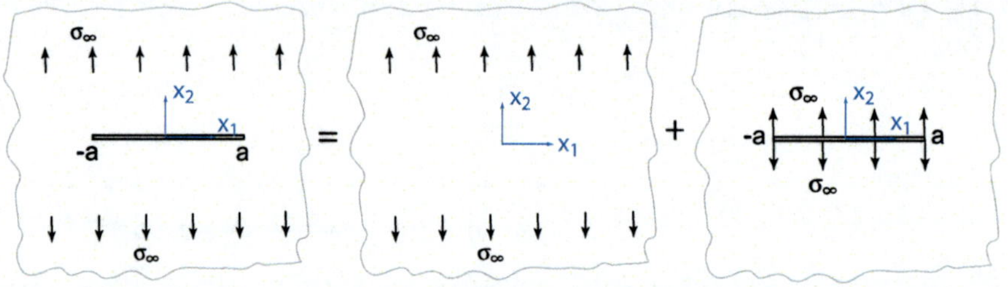

Fig. 2.7 Crack of length $2a$ in an infinite plate with far field stress $\sigma_{22} = \sigma_\infty$. Problem can be solved by superposition of uniform stress and crack in plate with no far field loading but with crack face pressures equal to σ_∞

Superposing the uniform far-field stress with the stress given by the stress functions, Eq. (2.69), yields $\sigma_1 = \mathrm{Re}\,\phi' - x_2\,\mathrm{Im}\,\phi''$, $\sigma_{22} = \mathrm{Re}\,\phi' + x_2\,\mathrm{Im}\,\phi'' + \sigma_\infty$ and $\sigma_{12} = -x_2\,\mathrm{Re}\,\phi''$. Substituting in ϕ' from Eq. (2.76) yields

$$\begin{aligned}
\sigma_{11} &= \sigma_\infty\left[\mathrm{Re}\left(\frac{z}{\sqrt{z^2-a^2}}\right) - x_2\,\mathrm{Im}\left(\frac{1}{\sqrt{z^2-a^2}} - \frac{z^2}{(z^2-a^2)^{3/2}}\right)\right] - \sigma_\infty,\\
\sigma_{22} &= \sigma_\infty\left[\mathrm{Re}\left(\frac{z}{\sqrt{z^2-a^2}}\right) + x_2\,\mathrm{Im}\left(\frac{1}{\sqrt{z^2-a^2}} - \frac{z^2}{(z^2-a^2)^{3/2}}\right)\right],\\
\sigma_{12} &= -\sigma_\infty x_2\,\mathrm{Re}\left(\frac{1}{\sqrt{z^2-a^2}} - \frac{z^2}{(z^2-a^2)^{3/2}}\right).
\end{aligned} \tag{2.78}$$

Evaluating the stresses along $z = x_1$:

$$\begin{aligned}
\sigma_{11}(x_1,0) &= \mathrm{Re}\left(\frac{\sigma_\infty x_1}{\sqrt{x_1^2-a^2}}\right) - \sigma_\infty,\\
\sigma_{22}(x_1,0) &= \mathrm{Re}\left(\frac{\sigma_\infty x_1}{\sqrt{x_1^2-a^2}}\right),\\
\sigma_{12}(x_1,0) &= 0.
\end{aligned}$$

Note that along the crack line, for $-a \le x_1 \le a$, $\mathrm{Re}(\frac{\sigma_\infty x_1}{\sqrt{x_1^2-a^2}}) = 0$, and hence $\sigma_{22} = 0$ as required, and $\sigma_{11} = -\sigma_\infty$. Along the crack the plate is in compression in the x_1 direction, which can lead to local buckling when large, thin, cracked sheets are loaded in tension. Using Eqs. (2.78) the stress fields σ_{11} and σ_{22}, normalized by σ_∞, are plotted in Fig. 2.8 for $0 \le x_1/a \le 2$, $0 \le x_2/a \le 1$.

We can determine the stress intensity factor by examining the solution near one of the crack tips. Let $r = x_1 - a$, $x_1 = r + a$ then as $r \to 0$, i.e. near the right hand crack tip

$$\sigma_{22}(r,0) = \frac{\sigma_\infty a}{\sqrt{r}\sqrt{2a}} = \frac{\sigma_\infty \sqrt{a}}{\sqrt{2r}}. \tag{2.79}$$

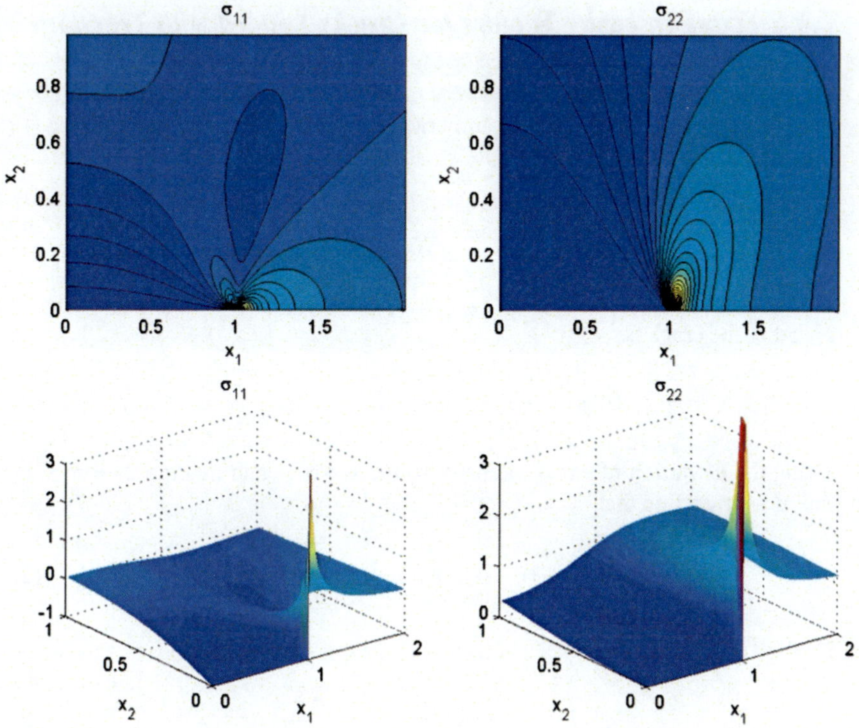

Fig. 2.8 Stress fields for finite crack in an infinite plate under tension. Stress normalized by σ_∞, coordinates normalized by a, from Eq. (2.78)

Using the definition of stress intensity factor, $K_I = \lim_{r\to 0} \sigma_{22}(r, 0)\sqrt{2\pi r}$ we find

$$K_I = \sigma_\infty \sqrt{\pi a}. \tag{2.80}$$

The opening displacement along the crack line can be found using Eq. (2.70) with ϕ' and ϕ as calculated above. The result is

$$u_2^+(x_1, 0) - u_2^-(x_1, 0) = \sigma_\infty \frac{\kappa + 1}{4\mu} \sqrt{a^2 - x_1^2}. \tag{2.81}$$

2.6.5 Stress Intensity Factor Under Remote Shear Loading

Similarly it can be shown that for a crack subject to remote stresses, $\sigma_{11} = 0$, $\sigma_{22} = 0$, $\sigma_{12} = \tau_\infty$ that the Mode-II stress intensity factor is

$$K_{II} = \tau_\infty \sqrt{\pi a}. \tag{2.82}$$

2.6.6 Stress Intensity Factors for Cracks Loaded with Tractions

We can develop equations for the stress intensity factors by focusing on the stresses near one crack tip. At the right hand crack tip, as $z \to a$, $z + a \approx 2a$, and $z - t \approx a - t$, hence from Eq. (2.73)

$$\phi' = \frac{1}{\pi}\frac{1}{\sqrt{z-a}}\frac{1}{\sqrt{2a}}\int_{-a}^{a} p_2(t)\sqrt{\frac{a+t}{a-t}}\,dt. \tag{2.83}$$

Using $\sigma_{22} = \mathrm{Re}\,\phi' + x_2\,\mathrm{Im}\,\phi'' + \sigma_\infty$ and specializing to $z = x_1$, $x_1 \geq a$, the stress ahead of the crack is

$$\sigma_{22}(x_1, 0) = \frac{1}{\pi\sqrt{2a}}\frac{1}{\sqrt{x_1 - a}}\int_{-a}^{a} p_2(t)\sqrt{\frac{a+t}{a-t}}\,dt + \sigma_\infty. \tag{2.84}$$

Using the definition of stress intensity factor as above and making the substitution $z = x_1 - a$ we find that

$$K_I = \frac{1}{\sqrt{\pi a}}\int_{-a}^{a} p_2(t)\sqrt{\frac{a+t}{a-t}}\,dt. \tag{2.85}$$

Similarly, K_{II} is given by

$$K_{II} = \frac{1}{\sqrt{\pi a}}\int_{-a}^{a} p_1(t)\sqrt{\frac{a+t}{a-t}}\,dt. \tag{2.86}$$

2.6.7 Asymptotic Mode I Field Derived from Full Field Solution

From Eqs. (2.85) and (2.83), as $z \to a$, ϕ' and ϕ'' for Mode-I loading may be written as

$$\begin{aligned} \phi' &= \frac{K_I}{\sqrt{2\pi}}\frac{1}{\sqrt{z-a}}, \\ \phi'' &= -\frac{K_I}{\sqrt{2\pi}}\frac{1}{2(z-a)^{3/2}}. \end{aligned} \tag{2.87}$$

Making the substitution $z - a = re^{i\theta}$, we have $1/\sqrt{z-a} = (1/\sqrt{r})e^{-i\theta/2} = (1/\sqrt{r})(\cos\theta/2 - i\sin\theta/2)$, and $1/(z-a)^{3/2} = (1/r^{3/2})e^{-3i\theta/2} = (1/r^{3/2})(\cos 3\theta/2 - i\sin 3\theta/2)$. Thus Eqs. (2.87) can be written as

$$\begin{aligned} \phi' &= \frac{K_I}{\sqrt{2\pi}}\frac{1}{\sqrt{r}}e^{-i\theta/2}, \\ \phi'' &= -\frac{K_I}{2\sqrt{2\pi}}\frac{1}{r^{3/2}}e^{-3i\theta/2}. \end{aligned} \tag{2.88}$$

Substituting the above into Eq. (2.69), and using $x_2 = r\sin\theta$ we obtain (using σ_{22} as an example)

$$\sigma_{22} = \frac{K_I}{\sqrt{2\pi}}\left(\frac{1}{\sqrt{r}}\cos\frac{\theta}{2} + r\ \sin\theta\frac{1}{2}\frac{1}{r^{3/2}}\sin\frac{3\theta}{2}\right).$$

Using the identity $\sin\theta = 2\sin\frac{\theta}{2}\cos\frac{\theta}{2}$, σ_{22} can be written as

$$\sigma_{22} = \frac{K_I}{\sqrt{2\pi r}}\cos\frac{\theta}{2}\left(1 + \sin\frac{\theta}{2}\sin\frac{3\theta}{2}\right). \tag{2.89}$$

The stress σ_{11} is the same but with a change in the sign of the second term,

$$\sigma_{11} = \frac{K_I}{\sqrt{2\pi r}}\cos\frac{\theta}{2}\left(1 - \sin\frac{\theta}{2}\sin\frac{3\theta}{2}\right). \tag{2.90}$$

In the same manner the shear stress is found to be

$$\sigma_{12} = \frac{K_I}{\sqrt{2\pi r}}\cos\frac{\theta}{2}\sin\frac{\theta}{2}\cos\frac{3\theta}{2}. \tag{2.91}$$

The asymptotic stress function, Eq. (2.87) can be integrated, yielding

$$\phi = \frac{K_I}{\sqrt{2\pi}}2\sqrt{z-a} = K_I\sqrt{\frac{2}{\pi}}\sqrt{r}e^{i\theta/2}. \tag{2.92}$$

Substituting Eq. (2.92) and Eq. (2.87) into Eqs. (2.70) we find (using u_2 as an example)

$$2\mu u_2 = \frac{\kappa+1}{2}K_I\sqrt{\frac{2}{\pi}}\sqrt{r}\sin\frac{\theta}{2} - r\sin\theta\frac{K_I}{\sqrt{2\pi r}}\cos\frac{\theta}{2}.$$

Collecting terms and using the identities $\sin\theta = 2\sin\frac{\theta}{2}\cos\frac{\theta}{2}$ and $2\cos^2\frac{\theta}{2} = 1 + \cos\theta$ the displacement can be written as

$$u_2 = \frac{K_I}{2\mu}\sqrt{\frac{r}{2\pi}}\sin\frac{\theta}{2}(\kappa - \cos\theta). \tag{2.93}$$

Similarly

$$u_1 = \frac{K_I}{2\mu}\sqrt{\frac{r}{2\pi}}\cos\frac{\theta}{2}(\kappa - \cos\theta). \tag{2.94}$$

2.6.8 Asymptotic Mode II Field Derived from Full Field Solution

Following similar procedures the Mode-II fields can be derived in Cartesian coordinates, resulting in the stress field,

$$\begin{pmatrix} \sigma_{11} \\ \sigma_{12} \\ \sigma_{22} \end{pmatrix} = \frac{K_{II}}{\sqrt{2\pi r}} \begin{pmatrix} -\sin\frac{\theta}{2}(2+\cos\frac{\theta}{2}\cos\frac{3\theta}{2}) \\ \cos\frac{\theta}{2}(1-\sin\frac{\theta}{2}\sin\frac{3\theta}{2}) \\ \sin\frac{\theta}{2}\cos\frac{\theta}{2}\cos\frac{3\theta}{2} \end{pmatrix}, \tag{2.95}$$

with displacements

$$\begin{pmatrix} u_1 \\ u_2 \end{pmatrix} = \frac{K_{II}}{2\mu}\sqrt{\frac{r}{2\pi}} \begin{pmatrix} \sin\frac{\theta}{2}(\kappa+2+\cos\theta) \\ -\cos\frac{\theta}{2}(\kappa-2+\cos\theta) \end{pmatrix}. \tag{2.96}$$

2.6.9 Stress Intensity Factors for Semi-infinite Crack

Consider a semi-infinite crack, Fig. 2.3, loaded with tractions over a region near the crack tip. In this case, in Eq. (2.85) let $-a \to -\infty$, $a+t \to 2a$, and transform coordinates by $a' = t - a$ so that

$$\begin{aligned} K_I &= \frac{1}{\sqrt{\pi}\sqrt{a}}\sqrt{2a}\int_{-\infty}^{0} \frac{p_2(a')}{\sqrt{-a'}}\,da', \\ K_I &= \sqrt{\frac{2}{\pi}}\int_{-\infty}^{0} \frac{p_2(a')}{\sqrt{-a'}}\,da'. \end{aligned} \tag{2.97}$$

Similarly,

$$K_{II} = \sqrt{\frac{2}{\pi}}\int_{-\infty}^{0} \frac{p_1(a')}{\sqrt{-a'}}\,da'. \tag{2.98}$$

2.7 Some Comments

How do we know that the solutions chosen above correspond to “Mode-I” and “Mode-II” as illustrated in Fig. 2.2. Perhaps the best manner to see this is to consider the displacement fields along the crack faces. Analyzing the Mode-I displacement field, Eq. (2.58), the reader can see that the relative motion of the crack faces is only in the x_2 direction, i.e. there is no relative sliding of the crack faces. Analyzing the Mode-II displacement field, Eq. (2.62) the reader will see that the crack faces do not open up, and that the top and bottom crack faces slide relative to each other. The lack of crack opening in Mode-II brings up questions regarding the effect of crack face friction on the growth of cracks loaded in Mode-II. Note also that if $K_I < 0$, Eq. (2.61) tell us that the crack faces will interpenetrate. As this is physically impossible it tells us that in such a case the crack faces will no longer be traction free, but will push against each other and effectively it will be as if there is no crack present. This would differ however, if in the unloaded state the crack had a finite opening, arising (for example) from corrosion or other effects.

Close enough to the crack tip the stress and displacement fields are completely determined by the values of K_I, K_{II}, and K_{III}. The various methods for determining these values in laboratory and real-world applications will be discussed later in this book.

What will happen if a crack is loaded in a way that both Mode-I and Mode-II are present? What if Mode-I, Mode-II and Mode-III are all three present? The crack tip stresses will be a superposition of the solutions above. The relative values of K_I, K_{II} and K_{III} will depend on the loading and on the geometry of the crack and of the cracked body.

2.7.1 Three-Dimensional Cracks

Of course we do not live in a 2D world. So what will be different in 3D? As a start consider the simple problem of an edge crack in a plate under tension as shown in Fig. 2.9. The crack front is straight through the thickness of the plate. The stress field details for this problem were studied using a multi-grid, 3D finite element analysis [9]. This is a pure mode-I problem. The results show that the in-plane stresses, σ_{11}, σ_{22} and σ_{12} are nearly constant through the thickness with the normal stresses dropping off by approximately 25% at the free surfaces. Thus the 2D stress fields provide an accurate description of the 3D problem.

However the out-of-plane stress, σ_{33} has considerable variation through the thickness. This is to be expected. In the center of the plate, very near the crack tip, the free surfaces appear to be infinitely far away relative to the distance to the crack front and thus it is expected that the stress state will be plane-strain in which the out-of-plane normal strain and stress are $\gamma_{33} = 0$ and $\sigma_{33} = \nu(\sigma_{11} + \sigma_{22})$. At the free surfaces plane stress conditions are expected with normal stress σ_{33} and normal strain $\gamma_{33} = -\nu(\gamma_{11} + \gamma_{22})$. Note that in the plane stress solution since γ_{33} will be singular, the out-of-plane displacement, u_3 would be infinite as $r \to \infty$! Not a physically realistic result. The variation of σ_{33} through the thickness along a line perpendicular to the plate and located 45° to the x_1 axis at different distances to the crack tip is shown in Fig. 2.10. The results show that in the center of the plate, very close to the crack tip, the stress field is plane strain. Further away from the crack, $r \approx 0.33B$ the field is plane stress. Very close to the crack tip plane strain predominates except in a boundary layer near the free surfaces.

What will the stresses be at the crack line for real cracks in three dimensional objects? The stresses will be given by a superposition of the Mode-I, Mode-II and Mode-III fields with the values of K_I, K_{II} and K_{III} varying at different locations

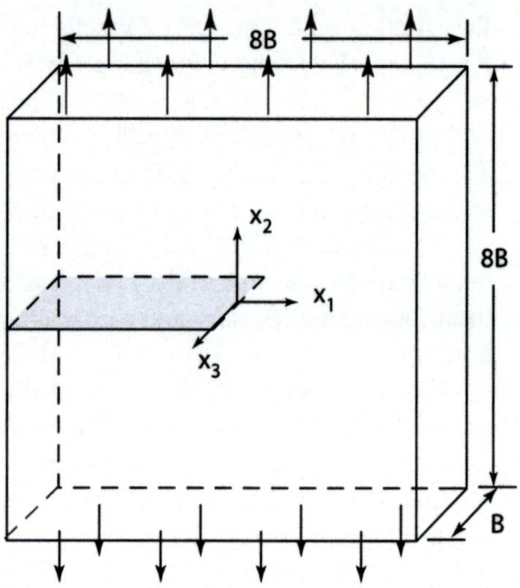

Fig. 2.9 Three-dimensional edge cracked plate loaded in tension. Adapted from [9]

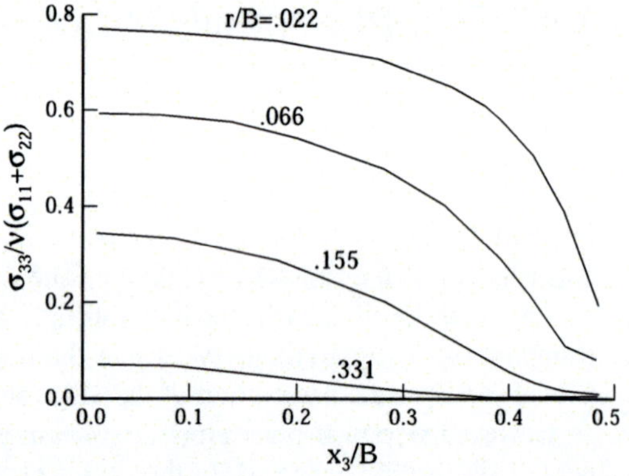

Fig. 2.10 Degree of plane strain through the plate thickness. Plots are for a line perpendicular to the plate located at $\theta = 45°$ and distances $r/B = 0.022, 0.066, 0.155, 0.331$ from the crack tip. $x_3/B = 0$ is the plate center. $x_3/B = 0.5$ is the free surface. Beyond $r \approx 0.33B$ the field is plane stress. Adapted from [9]

along the crack line. For example, consider the penny shaped crack of radius a shown in Fig. 2.11 subject to tension of σ_∞ at an angle of β to the crack surface. In this case the stress intensity factors are [10]

$$\begin{aligned} K_I &= \sigma_\infty \sqrt{\pi a} \frac{2}{\pi} \sin^2 \beta, \\ K_{II} &= \sigma_\infty \sqrt{\pi a} \frac{4}{\pi(2-\nu)} \sin\beta \cos\beta \cos\theta, \\ K_{III} &= -\sigma_\infty \sqrt{\pi a} \frac{4(1-\nu)}{\pi(2-\nu)} \sin\beta \cos\beta \sin\theta. \end{aligned} \tag{2.99}$$

At what point would the crack first begin to grow? And once growing, how would the fracture surface evolve? Such questions are still the topic of active research. To

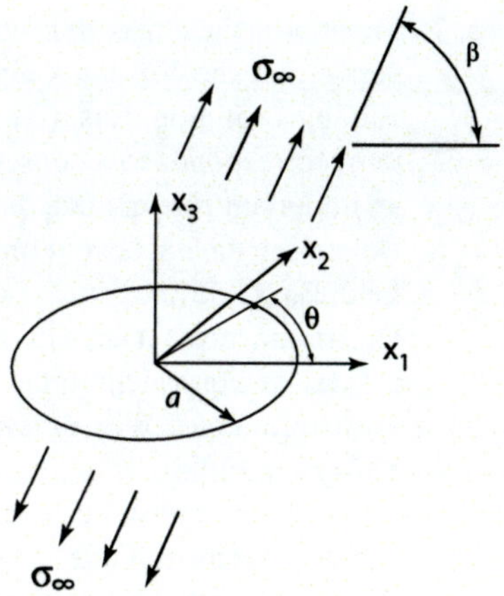

Fig. 2.11 Circular crack of radius a subject to uniform far field loading, σ_∞ at an angle of β to the crack surface. In the far-field, $\sigma_{33} = \sigma_\infty \sin^2\beta$, $\sigma_{11} = \sigma_\infty \cos^2\beta$, $\sigma_{13} = \sigma_\infty \sin\beta\cos\beta$. All other stress components are zero

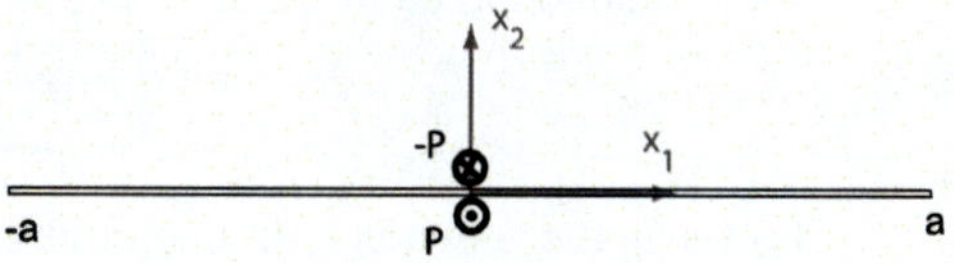

Fig. 2.12 Finite, anti-plane shear crack in an infinite body with line loads, $\pm P$ [F/L] applied

start to understand what will happen in such cases we need to study the energy flows in fracture and to address criteria for fracture. These follow in the next two chapters.

2.8 Exercises

1. Transform the Mode-III asymptotic stress field given in Eq. (2.23) into Cartesian coordinates using the appropriate coordinate transformation.
2. Using the procedure outlined in Sect. 2.4.1 determine the first term of the stress field for a crack of finite opening angle β. For what angle is the field no longer singular? Can you explain this on physical grounds?
3. Compare the asymptotic Mode-III crack tip field, Eq. (2.34) to the full-field solution Eq. (2.33) by plotting contours of the error incurred by approximating the full solution by the asymptotic solution. At approximately what distance from the crack tip does the error become greater than 10%?
4. Perform the same calculation as above for Mode-I, i.e. compare the error incurred by approximating the full-field solution for finite crack in tension with the asymptotic Mode-I field.
5. Consider a finite, anti-plane shear crack in an infinite body. Suppose that the crack is loaded by two equal and opposite line loads, P, acting on the center of the crack, as shown in Fig. 2.12. Using dimensional considerations determine how K_{III} scales with the load and crack length. Provide an intuitive explanation for this result.
6. Show that the stress function solution given in Eq. (2.43) satisfies the biharmonic equation.

7. Transform the stress field given in Eq. (2.49) into Cartesian components and verify that the Williams eigenfunction solution and the asymptotic complex variables solution, Eqs. (2.89)–(2.91) yield the same results. Show that the A_0 term corresponds to a constant stress parallel to the crack. This stress will play an important role in crack path stability and crack tip plasticity.
8. Determine the next two terms in the series solution for the Mode-I displacement field, Eq. (2.58).
9. Determine the next term in the series solution for the anti-symmetric stress field. Is there an equivalent to the constant stress term of magnitude A_0 found in the symmetric field? Why or why not?
10. Verify Eq. (2.86).
11. Using the complex variables method calculate the asymptotic Mode II stress and displacement fields.
12. Yet another way to find the asymptotic crack tip stress fields is to start with a stress function in the form of a Laurent series in z, i.e.

$$\phi = \sum A_n z^{\lambda_n}.$$

    (a) Using the above as a starting point in the Westergaard approach for Mode-I, calculate the first two terms $(r^{-1/2}, r^0)$ in the asymptotic stress field. (b) Do the same for the Mode-II problem.
13. Verify the integration of Eq. (2.75) leading to Eq. (2.76).
14. Verify that the stress function in Eq. (2.63) satisfies the biharmonic equation.
15. Verify Eq. (2.66) for the stress components $\sigma_{\alpha\beta}$.

References

1. C.Y. Hui, A. Ruina, Int. J. Fract. **72**, 97 (1995)
2. J.W. Brown, R.V. Churchill, *Complex Variables and Applications* (McGraw-Hill, New York, 2004)
3. M. Williams, J. Appl. Mech. **24**, 109 (1957)
4. S.P. Timoshenko, J.N. Goodier, *Theory of Elasticity* (McGraw-Hill, New York, 1969)
5. E.G. Coker, L.N.G. Filon, *A Treatise on Photoelasticity* (Cambridge University Press, Cambridge, 1931)
6. K. Hellan, *Introduction to Fracture Mechanics* (McGraw-Hill, New York, 1984)
7. H.M. Westergaard, J. Appl. Mech. **61**, A49 (1939)
8. L.I. Sedov, *A Course in Continuum Mechanics*, vol. 4 (Wolters-Noordhoff, Groningen, 1972)
9. I.D. Parsons, J.F. Hall, Eng. Fract. Mech. **33**, 45 (1989)
10. H. Tada, P.C. Paris, G.R. Irwin, *The Stress Analysis of Cracks Handbook* (ASME Press, New York, 2000)

Chapter 3
Energy Flows in Elastic Fracture

Abstract An alternative approach to understanding fracture is to look not to the stress fields but to the flow of energy into and out of an object that is susceptible to fracture. In this chapter the energy required for fracture will be considered and the connection between the energy flows and the crack tip stresses will be made. This connection also provides a powerful tool for the determination of stress intensity factors. Note that, except where noted, the energy approach to fracture is the same whether the material is linear or nonlinearly elastic.

3.1 Generalized Force and Displacement

To deal in general with the forces, displacements and energy flows in a solid body one would write all quantities in terms of volume and surface integrals. To simplify the notation, allowing us to concentrate on the fundamental ideas equivalent point forces and displacements will be defined. With this approach we need not consider the details of the loading and of the geometry. Two cases are considered. The first is the case of prescribed loading (traction and/or body force) and the second is the case of prescribed displacement.

3.1.1 Prescribed Loads

Assume that we have a solid body, acted on by a set of tractions and body forces. Assume that this body contains one or more cracks, and that the loads are held constant even if the cracks grow. The tractions $\mathbf{t}$ and the body force $\mathbf{b}$ could in this case be scaled by a scalar, P, such that $\mathbf{t} = P\hat{\mathbf{t}}$ and $\mathbf{b} = P\hat{\mathbf{b}}$. P is known as the generalized load with dimension [F]. If the displacements would change by $\delta\mathbf{u}$ then the work done by the forces on the body would be

$$\delta U = \int_{\mathscr{S}} \mathbf{t}\cdot\delta\mathbf{u}\,dS + \int_{\mathscr{V}} \mathbf{b}\cdot\delta\mathbf{u}\,dV = \int_{\mathscr{S}} P\hat{\mathbf{t}}\cdot\delta\mathbf{u}\,dS + \int_{\mathscr{V}} P\hat{\mathbf{b}}\cdot\delta\mathbf{u}\,dV. \quad (3.1)$$

A.T. Zehnder, *Fracture Mechanics*,
Lecture Notes in Applied and Computational Mechanics 62,
DOI 10.1007/978-94-007-2595-9_3, © Springer Science+Business Media B.V. 2012

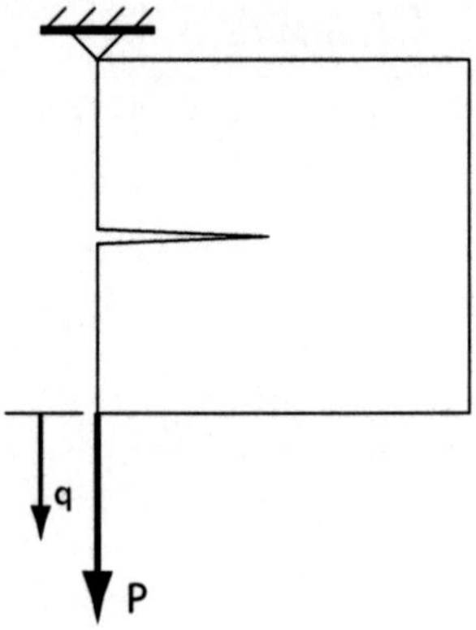

Fig. 3.1 Using the concept of generalized loads and displacements any loading may be represented as if it consisted of a single load, P with corresponding load point displacement, q

Since the loads remain fixed, the above could be rewritten as

$$\delta U = P\delta\left\{\int_{\mathscr{S}} \hat{\mathbf{t}}\cdot\mathbf{u}\,dS + \int_{\mathscr{V}} \hat{\mathbf{b}}\cdot\mathbf{u}\,dV\right\}.$$

This can be simplified by defining a generalized displacement, q, with dimension [L]

$$q \equiv \int_{\mathscr{S}} \hat{\mathbf{t}}\cdot\mathbf{u}\,dS + \int_{\mathscr{V}} \hat{\mathbf{b}}\cdot\mathbf{u}\,dV, \tag{3.2}$$

then the work done can be simply written as the product of a force and displacement increment,

$$\delta U = P\delta q. \tag{3.3}$$

Any loading applied to the body may now be represented by a single generalized force P with corresponding generalized displacement q as sketched in Fig. 3.1.

3.1.2 Prescribed Displacements

The generalized force and displacement may also be defined for the case of prescribed displacements. Scale the prescribed displacements by a $\mathbf{u} = q\hat{\mathbf{u}}$, where q is the generalized displacement with dimension [L]. Then the work done due to an increment of displacement $\delta\mathbf{u} = \delta q\hat{\mathbf{u}}$ is

$$\delta U = \int_{\mathscr{S}} \mathbf{t}\cdot\delta q\hat{\mathbf{u}}\,dS + \int_{\mathscr{V}} \mathbf{b}\cdot\delta q\hat{\mathbf{u}}\,dV,$$

$$\delta U = \delta q\left\{\int_{\mathscr{S}} \mathbf{t}\cdot\hat{\mathbf{u}}\,dS + \int_{\mathscr{V}} \mathbf{b}\cdot\hat{\mathbf{u}}\,dV\right\}.$$

Defining the generalized force, P by

$$P \equiv \int_{\mathscr{S}} \mathbf{t} \cdot \hat{\mathbf{u}} dS + \int_{\mathscr{V}} \mathbf{b} \cdot \hat{\mathbf{u}} dV, \tag{3.4}$$

the work increment is once again given by Eq. (3.3), $\delta U = P\delta q$.

3.2 Elastic Strain Energy

The elastic strain energy density is given by

$$W = \int_0^{\gamma} \sigma_{ij} d\gamma_{ij}. \tag{3.5}$$

Note that the stress can be found from

$$\sigma_{ij} = \frac{\partial W}{\partial \gamma_{ij}}. \tag{3.6}$$

The total strain energy in a body is

$$\Omega = \int_{\mathscr{V}} W dV. \tag{3.7}$$

The increment of strain energy density due to a displacement increment $\delta\mathbf{u}$ producing a strain increment $\delta\gamma$ is $\delta W = \sigma_{ij}\delta\gamma_{ij}$ and thus the increment in total strain energy is $\delta\Omega = \int_V \delta W \, dV$.

From Eqs. (3.1) and (3.3), in the absence of crack growth the work done on a solid due to a displacement increment is

$$\delta U = P\delta q = \int_{\mathscr{S}} \mathbf{t} \cdot \delta\mathbf{u} dS + \int_{\mathscr{V}} \mathbf{b} \cdot \delta\mathbf{u} dV.$$

Using indicial notation and replacing t_i by $\sigma_{ij} n_j$

$$\delta U = \int_{\mathscr{S}} \sigma_{ij} n_j \delta u_i \, dS + \int_{\mathscr{V}} b_i \delta u_i \, dV. \tag{3.8}$$

Applying the divergence theorem to the first of the above integrals we have

$$\delta U = \int_{\mathscr{V}} \left\{ \sigma_{ij},_j \delta u_i + \sigma_{ij} \delta u_i,_j + b_i \delta u_i \right\} dV.$$

Applying the equilibrium equation $\sigma_{ij},_j + b_i = 0$ and noting that $\sigma_{ij}\delta u_i,_j = \sigma_{ij}\delta\gamma_{ij} = \delta W$

$$\delta U = \int_{\mathscr{V}} \sigma_{ij} \delta\gamma_{ij} \, dV = \int_{\mathscr{V}} \delta W \, dV = \delta\Omega. \tag{3.9}$$

Thus we have shown that in the absence of crack growth, the increment of work done is equal to the increment in strain energy of the body.

For a prescribed displacement problem the total strain energy will be a function of the displacement, q and of the crack configuration, simplified here as being represented by crack area s. Thus we can write $\Omega = \Omega(q, s)$. If the crack does not grow, then the change in strain energy for an increment δq of displacement is

$$\delta\Omega = \frac{\partial\Omega}{\partial q}\delta q. \tag{3.10}$$

Knowing that (from Eq. (3.9)) $\delta\Omega = \delta U = P\delta q$ we can infer that

$$P = \frac{\partial\Omega}{\partial q}, \tag{3.11}$$

i.e. the generalized force is the derivative of the total strain energy with respect to the generalized displacement.

3.3 Energy Release Rate, G

To propagate a crack, energy must be supplied to the crack tip. This energy flows to the crack tip through the elasticity of the body and is dissipated via irreversible deformation, heat, sound and surface energy.

In the following we introduce the idea of energy release rate, G, the energy dissipated by fracture per unit new fracture surface area, ds. We will start with two specific cases, then develop the general definition of energy release rate.

3.3.1 Prescribed Displacement

If displacement is prescribed the strain energy, Ω can be considered as a function of applied displacement and the crack surface, i.e. $\Omega = \Omega(q, s)$. If the crack is allowed to propagate, increasing the fracture surface area by an amount δs, then the change in strain energy is

$$\delta\Omega = \frac{\partial\Omega}{\partial s}\delta s + \frac{\partial\Omega}{\partial q}\delta q. \tag{3.12}$$

As the crack grows, in the case of fixed (during crack growth) displacements $\delta q = 0$, thus

$$\delta\Omega = \frac{\partial\Omega}{\partial s}\delta s,$$

i.e. even though no external work is done on the body during crack growth, the strain energy changes in proportion to the increment of crack area. The energy change per unit area is called the energy release rate, G, with units of energy per area or [F·L/L^2], and is defined by

$$G = -\left.\frac{\partial \Omega}{\partial s}\right|_q, \tag{3.13}$$

with the $(\cdot)|_q$ notation to emphasize that the derivative is with respect to a fixed displacement. Written in terms of energy release rate the change in total strain energy for an increment δs of crack surface area is

$$\delta \Omega = -G\delta s. \tag{3.14}$$

It will be shown that G is always positive and hence that the body loses energy during crack growth. All of the energy dissipated during fracture flows to the crack from strain energy stored in the body prior to fracture.

3.3.2 Prescribed Loads

For the case of prescribed loading the increment in displacement, δq will not be zero during an increment of crack growth, hence the external loads will do work on the body as the crack grows. The displacement can be considered as a function of the load and fracture surface, i.e. $q = q(P, s)$ and the hence the strain energy is $\Omega = \Omega(q(P, s), s)$.

The increment in strain energy as the crack grows is

$$\delta \Omega = \frac{\partial \Omega}{\partial q}\frac{\partial q}{\partial P}\delta P + \frac{\partial \Omega}{\partial q}\frac{\partial q}{\partial s}\delta s + \left.\frac{\partial \Omega}{\partial s}\right|_q \delta s. \tag{3.15}$$

Since the force is fixed, $\delta P = 0$. Substituting $\frac{\partial \Omega}{\partial q} = P$ and $\frac{\partial q}{\partial s}\delta s = \delta q$, and noting that $\frac{\partial \Omega}{\partial s}|_q = -G$, the change in strain energy can be written as

$$\delta \Omega = P\delta q - G\delta s, \tag{3.16}$$

i.e. *the change in stored energy equals the energy input minus the energy dissipated by fracture.* The above can be rewritten as

$$\delta \Omega = \delta(Pq) - \delta P q - G\delta s.$$

Noting that $\delta P = 0$ in a prescribed loading problem,

$$\delta(\Omega - Pq) = -G\delta s,$$

from which it can be inferred that for fixed loads

$$G = -\frac{\partial}{\partial s}(\Omega - Pq). \tag{3.17}$$

3.3.3 General Loading

Using the definitions of the generalized force and displacement for the case of prescribed loading, and noting that in this case $\mathscr{S} = \mathscr{S}_t$, Eq. (3.17) is

$$G = -\frac{\partial}{\partial s}\left[\Omega - P\left\{\int_{\mathscr{S}} \hat{\mathbf{t}} \cdot \mathbf{u}\, dS + \int_{\mathscr{V}} \hat{\mathbf{b}} \cdot \mathbf{u}\, dV\right\}\right],$$

$$= -\frac{\partial}{\partial s}\left[\Omega - \left\{\int_{\mathscr{S}_t} \mathbf{t} \cdot \mathbf{u}\, dS + \int_{\mathscr{V}} \mathbf{b} \cdot \mathbf{u}\, dV\right\}\right].$$

The quantity in square brackets is the potential energy.

For the case of prescribed displacement, noting in this case that $\mathbf{b} = \mathbf{0}$ and $\mathscr{S} = \mathscr{S}_u$ (or $\int_{\mathscr{S}_t} (\cdot)\, dS = 0$), Eq. (3.13) can be written as

$$G = -\frac{\partial \Omega}{\partial s} = -\frac{\partial}{\partial s}\left[\Omega - \left\{\int_{\mathscr{S}_t} \mathbf{t} \cdot \mathbf{u}\, dS + \int_{\mathscr{V}} \mathbf{b} \cdot \mathbf{u}\, dV\right\}\right].$$

Note that the integrals equal zero in the prescribed displacement case.

Denoting the potential energy by Π,

$$\Pi = \Omega - \left\{\int_{\mathscr{S}_t} \mathbf{t} \cdot \mathbf{u}\, dS + \int_{\mathscr{V}} \mathbf{b} \cdot \mathbf{u}\, dV\right\}, \tag{3.18}$$

where the total strain energy, Ω, is defined in Eq. (3.7), the energy release rate can be written in a common form as

$$G \equiv -\frac{\partial \Pi}{\partial s}, \tag{3.19}$$

i.e. energy release rate is the change of potential energy per unit crack area. This is taken to be the fundamental definition of G. Any other loading, for example loading by a spring or with mixed boundary conditions (load prescribed on part of the boundary and displacement on other parts) will fall in between the two extreme cases of prescribed loading or prescribed displacement. Further interpretation, generalization and application of the energy release rate will be discussed in this and in subsequent chapters. You'll not escape G; energy release rate is a ubiquitous concept in fracture mechanics.

3.4 Interpretation of G from Load-Displacement Records

3.4.1 Multiple Specimen Method for Nonlinear Materials

Suppose that we can perform several experiments on an elastic, cracked body. In the experiment the load, P and the load-point displacement, q are measured and recorded.

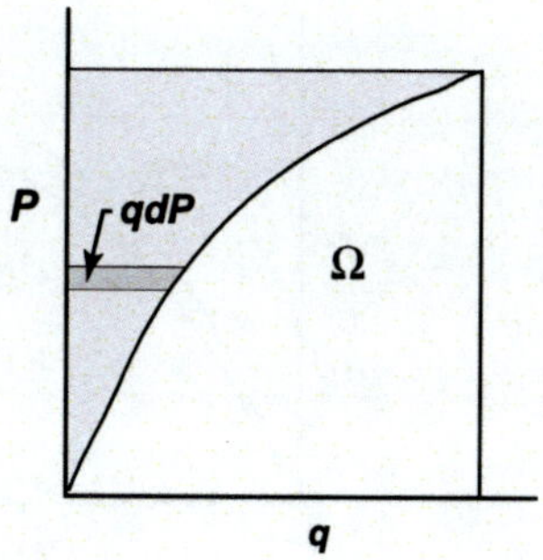

Fig. 3.2 Strain energy, Ω and complementary energy (*shaded area*)

Since the material is elastic there is a specific relation between P and q and one can write the strain energy as $\Omega = \Omega(P, s)$ for prescribed loads or as $\Omega = \Omega(q, s)$ for prescribed displacements.

In the prescribed displacement case, $\Omega = \Omega(q, s)$, and from Eqs. (3.10) and (3.13)

$$G = -\frac{\partial \Omega}{\partial s},$$

$$P = \frac{\partial \Omega}{\partial q}.$$

Taking derivatives of G and P,

$$-\frac{\partial G}{\partial q} = \frac{\partial P}{\partial s}.$$

The above can be integrated with respect to q, holding the crack length fixed

$$G = -\int_0^q \frac{\partial P}{\partial s} dq. \tag{3.20}$$

An alternative integral expression for G may be obtained by considering the prescribed load case in which $\Omega = \Omega(P, s)$. However, first an intermediate result is needed. Consider the complementary energy, $Pq - \Omega$, shown as the shaded area in Fig. 3.2. This energy is given by

$$Pq - \Omega = \int_o^P q(P', s) dP',$$

thus

$$d(Pq - \Omega) = q dP,$$

or

$$q = -\frac{\partial}{\partial P}(\Omega - Pq). \tag{3.21}$$

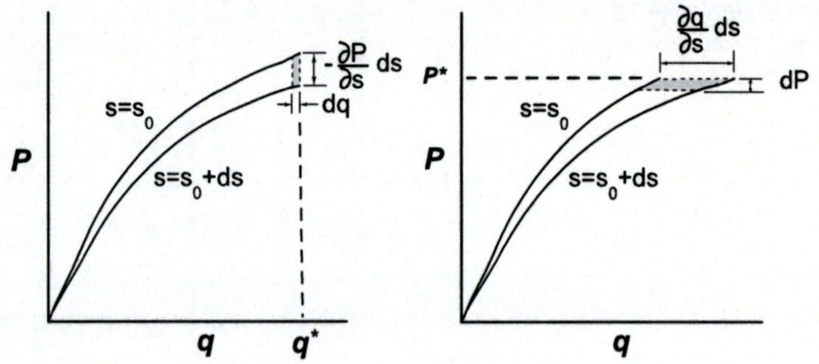

Fig. 3.3 Interpretation of Eqs. (3.20) and (3.22)

From Eq. (3.17)

$$G = -\frac{\partial}{\partial s}(\Omega - Pq).$$

Taking derivatives

$$\frac{\partial G}{\partial P} = \frac{\partial q}{\partial s}.$$

Integrating with respect to P

$$G = \int_0^P \frac{\partial q}{\partial s} dP. \tag{3.22}$$

Equations (3.20) and (3.22) provide integrals that can be evaluated to determine the energy release rate. The interpretation of these integrals is shown in Fig. 3.3.

To understand Eq. (3.20) consider loading a body with crack area s_0 up to a displacement of q^* as shown. At this displacement, allow the crack to grow a small amount, ds while holding the displacement fixed. Due to crack growth the body will be more compliant (or less stiff) and hence the load will drop. Now unload the body and the displacement will return to zero since the body is assumed to be elastic. At every value of q the difference in P between the loading curve for crack area s and unloading curve for crack area $s + ds$ is approximately $-\frac{\partial P}{\partial s} ds$. Hence the area between the curves is approximately

$$Gds = -ds \int_0^q \frac{\partial P}{\partial s} dq = \text{area between curves.}$$

To understand Eq. (3.22) consider loading a body with crack area s up to a load P^* as shown in Fig. 3.3. Holding the load fixed, allow the crack to grow. Since the body is now more compliant the displacement will increase. Now unload to zero load. The displacement difference between the two curves is approximately $\frac{\partial q}{\partial s} da$. Thus the area between the two curves is

$$G\,ds = ds \int_0^P \frac{\partial q}{\partial s} dP = \text{area between curves.}$$

Both of these results have the very simple interpretation that the difference between the work done on the sample and the energy that can be recovered upon unloading is equal to the energy release rate, G multiplied by the increment in crack area, ds.

In principle, one could construct $G = G(P, s)$, or $G = G(q, s)$ by performing a series of experiments on a set of samples identical except for different crack lengths. The derivatives $\frac{\partial P}{\partial s}$ or $\frac{\partial q}{\partial s}$ could be determined by differentiating the P vs. q data with respect to crack length. This so-called multiple specimen technique for determination of G has been used, but has largely been superseded by other techniques and by computational methods.

3.4.2 Compliance Method for Linearly Elastic Materials

When the material is *linearly* elastic the P vs. q curve is linear and the analysis of G can be greatly simplified.

The displacement per unit force applied is called the compliance, C with units of [L/F]. It is the inverse of the slope of the P vs. q curve,

$$C = \frac{q}{P}. \tag{3.23}$$

The strain energy is given by

$$\Omega = \frac{1}{2}Pq = \frac{1}{2}\frac{q^2}{C} = \frac{1}{2}P^2C \tag{3.24}$$

Recall from Eq. (3.13) that for fixed displacement $G = -\frac{\partial \Omega}{\partial s}$. Replacing Ω by $\frac{1}{2}\frac{q^2}{C}$

$$G = -\frac{\partial}{\partial s}\left(\frac{1}{2}\frac{q^2}{C}\right) = \frac{1}{2}\frac{q^2}{C^2}\frac{\partial C}{\partial s} = \frac{1}{2}P^2\frac{\partial C}{\partial s}.$$

Recall from Eq. (3.17) that for fixed load $G = -\frac{\partial}{\partial s}(\Omega - Pq)$. Replacing Ω by $\frac{1}{2}P^2C$ and noting that $Pq = P^2C$, the quantity $\Omega - Pq = -\frac{1}{2}P^2C$ and

$$G = -\frac{\partial}{\partial s}\left(-\frac{1}{2}P^2C\right) = \frac{1}{2}P^2\frac{\partial C}{\partial s}. \tag{3.25}$$

Thus the equation for G does not depend on whether loads or displacements are fixed. However, for fixed displacement G is a decreasing function of crack length, while for fixed load G increases with crack length.

If the crack grows under fixed displacement then as discussed above all of the energy required for fracture comes from strain energy in the body, i.e. $Gds = -\frac{\partial \Omega}{\partial s}$. In the case of fixed force, the applied loads do work on the body during crack growth. In this case, from Eq. (3.17), the energy balance is

$$\begin{aligned} G &= -\frac{\partial}{\partial s}(\Omega - Pq) \\ &= -\frac{\partial \Omega}{\partial s} + \frac{\partial}{\partial s}(Pq), \\ G + \frac{\partial \Omega}{\partial s} &= P\frac{\partial q}{\partial s} = P^2\frac{\partial C}{\partial s}, \\ G + \frac{\partial \Omega}{\partial s} &= 2G \quad \Rightarrow \quad G = \frac{\partial \Omega}{\partial s}. \end{aligned}$$

Thus it is seen that if the crack grows under fixed load conditions $2G\,ds$ units of work are done on the body. This work is split evenly between increasing the strain energy and energy loss due to crack extension.

3.4.3 Applications of the Compliance Method

3.4.3.1 Determination of *G* in DCB Sample

In many test specimens and applications one can perform a stress analysis to determine the load-point displacement as a function of crack area. This analysis could be analytical or computational. Consider, for example the test specimen shown in Fig. 3.4. This geometry is known as the double cantilever beam, or DCB. If $a \gg h$ then the part of the sample to the left of the crack tip can be considered to consist of two cantilevered beams. The portion of the sample to the right of the crack tip is approximated as stress free and hence as having no displacement. (This is reasonable except just at the base of the "beams" where there is some compliance.) The displacement of a single beam is $v_{max} = Pa^3/(3EI)$, where I is the moment of inertia, given by $I = Bh^3/12$. In the DCB specimen, $q = 2v_{max}$ and hence $q = 8Pa^3/(EBh^3)$. Thus the compliance is

$$C = \frac{q}{P} = \frac{8a^3}{EBh^3}.$$

Note that the change fracture surface area $ds = Bda$, thus $\frac{\partial C}{\partial s} = \frac{\partial C}{\partial a}\frac{da}{ds} = \frac{\partial C}{\partial a}\frac{1}{B}$. The derivative is

$$\frac{\partial C}{\partial a} = \frac{24a^2}{EBh^3}.$$

Hence using the above and Eq. (3.25) we can write an expression for $G(P, a)$,

$$G(P, a) = \frac{P^2}{B^2}\frac{12a^2}{Eh^3}. \tag{3.26}$$

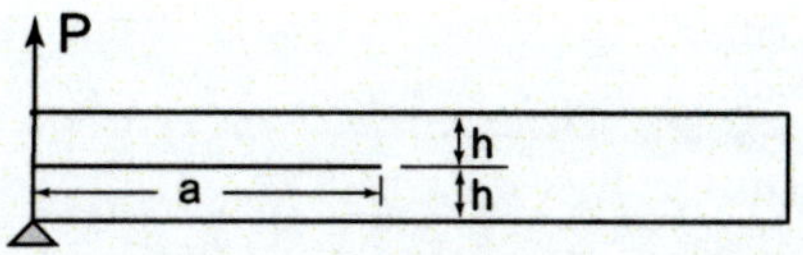

Fig. 3.4 DCB geometry, thickness B, load-point displacement q. This geometry is particularly useful for the study of delamination in composite materials

As a function of the applied displacement, q,

$$G(q,a)=\frac{3}{16}\frac{Eh^3q^2}{a^4}. \tag{3.27}$$

Thus G is an increasing function of a for a fixed load, but a decreasing function of a for a fixed displacement.

3.4.3.2 Use of Compliance to Determine Crack Length

Under cyclic loading a crack will generally grow a small amount per cycle of loading. If by experimental, analytical or computational analysis one can pre-determine $C(s)$ for a particular test specimen, then if during the experiment the compliance is measured then the crack length can be determined from the inverse of $C(s)$. Compliance is determined experimentally by measuring the load and load-point displacement simultaneously. Load point displacement can be measured using an LVDT or clip gauge, see the experimental section for further details.

3.5 Crack Closure Integral for G

Recall that the energy release rate is defined as $G=-\frac{\partial \Pi}{\partial a}$. By considering the energy differences between a body before and after crack growth it will be shown that G can be written in terms of the stresses and displacements at the crack tip, and hence that for elastic materials, G and $\{K_I, K_{II}, K_{III}\}$ are related.

Equations (3.8) and (3.9) can be combined to state the "principle of virtual work,"

$$\int_{\mathscr{S}}\sigma_{ij}n_j\delta u_i\,dS+\int_{\mathscr{V}}b_i\delta u_i\,dV=\int_{\mathscr{V}}\sigma_{ij}\delta\gamma_{ij}\,dV=\int_{\mathscr{V}}\delta W\,dV. \tag{3.28}$$

Following the derivation given by Rice [1] consider a body containing a cavity as shown in Fig. 3.5. The body is loaded with prescribed tractions and displacements, $\mathbf{t}=\mathbf{t}^0$ on $\mathscr{S}_t$ and $\mathbf{u}=\mathbf{u}^0$ on $\mathscr{S}_u$. The solution to the associated boundary value problem of linear elasticity is $\{\sigma^0,\gamma^0,\mathbf{u}^0\}$ on $\mathscr{V}$. Keeping the same boundary conditions, allow the cavity to grow so that now the volume is $\mathscr{V}-\Delta\mathscr{V}$ and the surface is $\mathscr{S}+\Delta\mathscr{S}$, where ΔV is the volume removed and $\Delta\mathscr{S}$ is the new, traction free

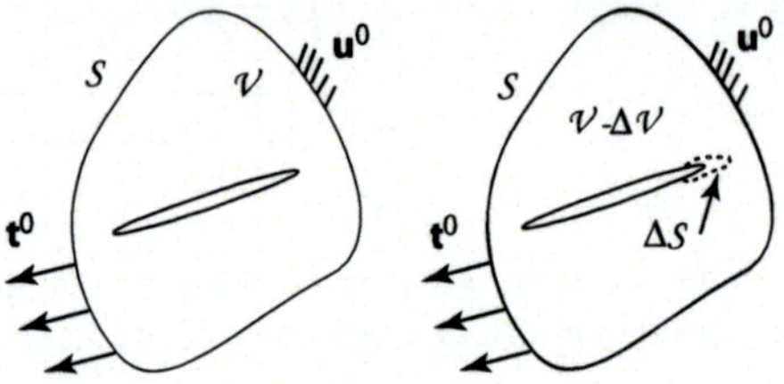

Fig. 3.5 Body $\mathscr{V}$ and surface $\mathscr{S}$. $\mathscr{S}$ includes surface of cavity. After the cavity grows, body is $\mathscr{V} - \Delta\mathscr{V}$ and surface is $\mathscr{S} + \Delta\mathscr{S}$. Tractions $\mathbf{t}^0$ are prescribed on $\mathscr{S}_t$. Displacements $\mathbf{u}^0$ are prescribed on $\mathscr{S}_u$

surface created by allowing the cavity to grow. The solution to the new problem is $\{\sigma^0 + \Delta\sigma,\ \gamma^0 + \Delta\gamma, \mathbf{u}^0 + \Delta\mathbf{u}\}$ on $\mathscr{V} - \Delta V$.

Writing the strain energy density W as a function of the strain, γ, the difference between the potential energy before and after growth of the cavity is

$$-\Delta\Pi = \int_{\mathscr{V}} W(\gamma^0)\,dV - \int_{\mathscr{S}_t} t_i^0 u_i^0\,dS - \int_{\mathscr{V}-\Delta\mathscr{V}} W(\gamma^0 + \Delta\gamma)\,dV$$
$$+ \int_{\mathscr{S}_t+\Delta\mathscr{S}} (t_i^0 + \Delta t_i)(u_i^0 + \Delta u_i)\,dS. \tag{3.29}$$

Noting that $\Delta t_i^0 = 0$ on $\mathscr{S}_t$, $\Delta u_i = 0$ on $\mathscr{S}_u$, $t_i^0 + \Delta t_i^0 = 0$ on $\Delta\mathscr{S}$,

$$\int_{\mathscr{S}_t+\Delta\mathscr{S}} (t_i^0 + \Delta t_i)(u_i^0 + \Delta u_i)\,dS = \int_{\mathscr{S}_t} t_i^0 (u_i^0 + \Delta u_i^0)\,dS, \tag{3.30}$$

and applying the principle of virtual work

$$\int_{\mathscr{S}_t} t_i^0 \Delta u_i\,dS = \int_{\mathscr{S}+\Delta\mathscr{S}} (t_i^0 + \Delta t_i)\Delta u_i\,dS = \int_{\mathscr{V}-\Delta V} (\sigma_{ij}^0 + \Delta\sigma_{ij})\Delta\gamma_{ij}. \tag{3.31}$$

Substituting Eq. (3.31) into Eq. (3.29), simplifying and re-arranging

$$-\Delta\Pi = \int_{\Delta\mathscr{V}} W(\gamma^0)\,dV$$
$$+ \int_{\mathscr{V}-\Delta\mathscr{V}} \left\{(\sigma_{ij}^0 + \Delta\sigma_{ij})\Delta\gamma_{ij} - \left[W(\gamma^0 - \Delta\gamma) - W(\gamma^0)\right]\right\} dV. \tag{3.32}$$

To further simplify Eq. (3.32) two side calculations are needed.

Side calculation 1: From the definition of W recall that $W(\gamma^0) = \int_0^{\gamma^0} \sigma_{ij}(\gamma)d\gamma_{ij}$. Recalling as well $\sigma_{ij}d\gamma_{ij} = \sigma_{ij}du_{i,j}$, the term

$$\int_{\mathscr{V}-\Delta\mathscr{V}} \left[W(\gamma^0 + \Delta\gamma) - W(\gamma^0)\right] dV$$
$$= \int_{\mathscr{V}-\Delta\mathscr{V}} \left\{\int_0^{\gamma^0+\Delta\gamma} \sigma_{ij}d\gamma_{ij} - \int_0^{\gamma^0} \sigma_{ij}d\gamma_{ij}\right\} dV$$
$$= \int_{\mathscr{V}-\Delta\mathscr{V}} \int_{\gamma^0}^{\gamma^0+\Delta\gamma} \sigma_{ij}d\gamma_{ij}\,dV$$

$$= \int_{\mathscr{V}-\Delta\mathscr{V}} \int_{u_{i,j}^0}^{u_{i,j}^0+\Delta u_{i,j}} \sigma_{ij} du_{i,j}\, dV$$

$$= \int_{\mathscr{V}-\Delta\mathscr{V}} \int_{u_{i,j}^0}^{u_{i,j}^0+\Delta u_{i,j}} (\sigma_{ij} du_i)_{,j}\, dV \quad (\text{using } \sigma_{ij,j} = 0).$$

Applying the divergence theorem

$$\int_{\mathscr{V}-\Delta\mathscr{V}} [W(\gamma^0 + \Delta\gamma) - W(\gamma^0)]\, dV = \int_{\mathscr{S}+\Delta\mathscr{S}} \int_{u_i^0}^{u_i^0+\Delta u_i} \sigma_{ij}\, du_i n_j\, dS$$

$$= \int_{\mathscr{S}+\Delta\mathscr{S}} \int_{u_i^0}^{u_i^0+\Delta u_i} t_i\, du_i\, dS. \qquad (3.33)$$

Side calculation 2:

$$\int_{\mathscr{V}-\Delta\mathscr{V}} (\sigma_{ij}^0 + \Delta\sigma_{ij})\Delta\gamma_{ij}\, dV = \int_{\mathscr{V}-\Delta\mathscr{V}} \big((\sigma_{ij}^0 + \Delta\sigma_{ij})\Delta u_i\big)_{,j}\, dV$$

$$= \int_{\mathscr{S}+\Delta\mathscr{S}} (t_i^0 + \Delta t_i)\Delta u_i\, dS. \qquad (3.34)$$

Substituting Eqs. (3.33) and (3.34) into Eq. (3.32),

$$-\Delta\Pi = \int_{\Delta\mathscr{V}} W(\gamma^0)\, dV$$

$$+ \int_{\mathscr{S}+\Delta\mathscr{S}} \left\{ (t_i^0 + \Delta t_i)\Delta u_i - \int_{u_i^0}^{u_i^0+\Delta u_i} t_i du_i \right\} dS. \qquad (3.35)$$

Noting that $\mathscr{S} + \Delta\mathscr{S} = \mathscr{S}_t \bigcup \mathscr{S}_u \bigcup \Delta\mathscr{S}$, $\Delta t_i = 0$ on $\mathscr{S}_t$, $\Delta u_i = 0$ on $\mathscr{S}_u$, and $t_i^0 + \Delta t_i = 0$ on $\Delta\mathscr{S}$, one finds $\int_{\mathscr{S}+\Delta\mathscr{S}} (t_i^0 + \Delta t_i)\Delta u_i\, dS = \int_{\mathscr{S}_t} t_i^0 \Delta u_i\, dS$. Similarly, $\int_{\mathscr{S}+\Delta\mathscr{S}} \int_{u_i^0}^{u_i^0+\Delta u_i} t_i du_i\, dS = \int_{\mathscr{S}_t} \int_{u_i^0}^{u_i^0+\Delta u_i} t_i du_i\, dS + \int_{\Delta\mathscr{S}} \int_{u_i^0}^{u_i^0+\Delta u_i} t_i du_i\, dS$. On $\mathscr{S}_t$, $t_i = t_i^0$, thus the $\int_{\mathscr{S}_t}$ term in the above is $\int_{\mathscr{S}_t} t_i^0 \Delta u_i\, dS$. Substituting into Eq. (3.35) we have

$$-\Delta\Pi = -\int_{\Delta\mathscr{S}} \int_{u_i^0}^{u_i^0+\Delta u_i} t_i du_i\, dS + \int_{\Delta\mathscr{V}} W(\gamma^0)\, dV. \qquad (3.36)$$

Equation (3.36) has the interpretation that *change in potential energy = work done in releasing tractions on $\Delta\mathscr{S}$ + strain energy of removed material, $\delta\mathscr{V}$.*

If instead of a cavity we have a crack and we allow that crack to grow, creating new surface $\Delta\mathscr{S}$ then the volume removed will be zero and the change in potential

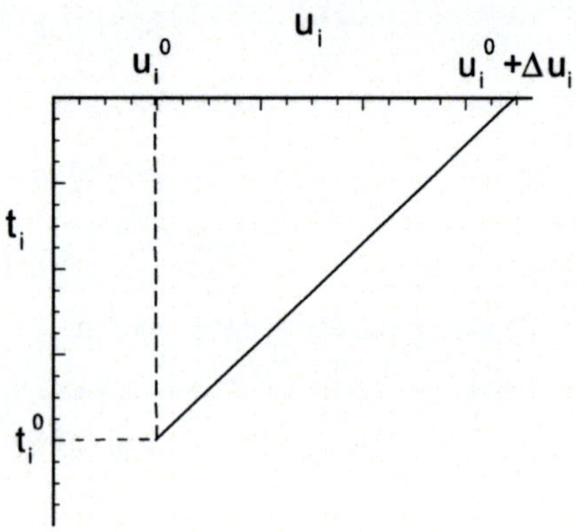

Fig. 3.6 Schematic of traction and displacement ahead of crack, before (u_i^0, t_i^0) and after crack growth $(u_i^0 + \Delta u_i, 0)$ in a linearly elastic material

energy will be

$$-\Delta \Pi = -\int_{\Delta \mathscr{S}} \int_{u_i^0}^{u_i^0+\Delta u_i} t_i du_i \, dS.$$

Some explanation of the above is in order. The traction will depend on the stress, which in turn depends on the strain, and hence on the gradient of the displacement field. Thus $t_i = t_i(\nabla \mathbf{u})$. Ahead of the new crack surface, $u_i = u_i^0$ and $t_i = t_i^0$. When the crack grows, the displacement is $u_i^0 + \Delta u_i$ and the traction drops to zero.

In a *linearly elastic material*, as sketched in Fig. 3.6, $\int_{u_i^0}^{u_i^0+\Delta u_i} t_i du_i = \frac{1}{2} t_i^0 \Delta u_i$, thus

$$-\Delta \Pi = -\int_{\Delta \mathscr{S}} \frac{1}{2} t_i^0 \Delta u_i \, dS.$$

The total new surface area created in growing the void (Fig. 3.5) is $\Delta \mathscr{S}$. Taking the void to be a crack, the new *fracture surface area*, Δs is half of $\Delta \mathscr{S}$. Another way to think of it is that let Δs^+ and Δs^- be the top and bottom fracture surfaces, then $\Delta \mathscr{S} = \Delta s^+ + \Delta s^-$. By the definition of energy release rate (Eq. (3.19))

$$G = -\frac{\partial \Pi}{\partial s} = \lim_{\Delta s \to 0} -\frac{1}{\Delta s} \left(\int_{\Delta s^+} \frac{1}{2} t_i^0 \Delta u_i \, dS + \int_{\Delta s^-} \frac{1}{2} t_i^0 \Delta u_i \, dS \right).$$

Noting that the tractions, t_i^+ on Δs^+ and t_i^- on Δs^- are equal and opposite, i.e. $t_i^- = -t_i^+$, then taking Δs to be Δs^+ we have

$$G = -\frac{\partial \Pi}{\partial s} = \lim_{\Delta s \to 0} -\frac{1}{\Delta s} \int_{\Delta s} \frac{1}{2} t_i^0 (\Delta u_i^+ - \Delta u_i^-) \, dS, \qquad (3.37)$$

where Δu_i^+ and Δu_i^- are the displacement increments evaluated on Δs^+ and Δs^- respectively and t_i^0 is the traction released on Δs^+. This result is valid for *any* elastic material and for any direction of crack growth. The above is known as the crack closure integral and was first given by Irwin [2]. Note that $\Delta u_i^+ - \Delta u_i^-$ is the crack opening (or sliding for Mode-II and Mode-III) displacement.

What is the sign of G? Considering the case of a tensile crack, the traction on Δs^+ would be downwards (negative), while the crack opening $\Delta u_i^+ - \Delta u_i^-$ would be positive, thus G would be positive.

3.6 G in Terms of K_I, K_{II}, K_{III} for 2D Cracks That Grow Straight Ahead

Although Eq. (3.37) applies for any direction or pattern of crack growth, calculating the actual value of G is difficult except for the case in which a crack grows straight ahead in a 2D body, for example, the crack in Fig. 3.4 with length a grows to a new length $a + \Delta a$ along the line $x_2 = 0$. In this case the tractions and displacements ahead of the crack can both be found from Eqs. (2.23), (2.49), and (2.58).

With respect to the coordinate system of Fig. 2.3, and for a crack increment of Δa, $\Delta s = B\Delta a$ and the integral $\int_{\Delta s}(\cdot)ds$ will be $B\int_{\Delta a}(\cdot)\,dx_1$.

For a crack that grows straight ahead $\Delta u_i^- = -\Delta u_i^+$. Hence the crack opening displacement $\Delta u_i^+ - \Delta u_i^- = 2\Delta u_i^+$, and $\Delta u_i^+ = u_i(\Delta a - x_1, \pi)$ where u_i is the crack tip displacement field (subtracted from any displacement existing along the crack line prior to crack growth) in polar coordinates. Along Δs^+, $\mathbf{n} = -1\mathbf{e}_2$, thus $t_i^0 = -\sigma_{i2}(x_1, 0)$ where σ_{i2} is the crack tip stress field in polar coordinates. The crack closure integral, Eq. (3.37), may now be written as

$$G = \lim_{\Delta a \to 0} \frac{1}{\Delta a} \int_0^{\Delta a} \sigma_{i2}(x_1, 0) u_i(\Delta a - x_1, \pi) dx_1. \tag{3.38}$$

3.6.1 *Mode-III Loading*

In a Mode-III crack, since $\sigma_{12} = \sigma_{22} = 0$, and $\sigma_{32}(x_1, 0) = \frac{K_{III}}{\sqrt{2\pi x_1}}$ the only non-zero term in Eq. (3.38) is for $i = 3$. The displacement is $u_3(\Delta a - x_1, \pi) = \sqrt{\frac{2}{\pi}}\frac{K_{III}}{\mu}(\Delta a - x_1)^{1/2}$. Thus

$$\begin{aligned} G &= \lim_{\Delta a \to 0} \frac{1}{\Delta a} \int_0^{\Delta a} \frac{K_{III}}{\sqrt{2\pi x_1}} \sqrt{\frac{2}{\pi}} \frac{K_{III}}{\mu} (\Delta a - x_1)^{1/2} dx_1 \\ &= \lim_{\Delta a \to 0} \frac{1}{\Delta a} \frac{K_{III}^2}{\mu\pi} \int_0^{\Delta a} \sqrt{\frac{\Delta a - x_1}{x_1}} dx_1. \end{aligned}$$

Noting that $\int_0^{\Delta a} \sqrt{\frac{\Delta a - x_1}{x_1}} dx_1 = \frac{\Delta a \pi}{2}$,

$$G = \frac{K_{III}^2}{2\mu}. \tag{3.39}$$

Thus G and K_{III} are in this case equivalent. A check will confirm that the units of G are [F/L] = [F·L/L^2], or energy per area.

Note that although Eq. (3.39) is valid only if the crack grows straight ahead, it could be used to calculate K_{III} no matter where the crack grows. How is this so? Suppose the crack configuration and loading are given and suppose further that you have a way to calculate G for straight ahead crack growth. Equation (3.39) could then be applied to determine K_{III}. Were the crack to propagate, it might grow in some other direction, in which case the *actual* energy release rate will not be given by Eq. (3.39). This, however, does not invalidate the use of Eq. (3.39) for calculating K_{III}.

3.6.2 Mode I Loading

For a crack under tensile, or Mode-I loading, ahead of the crack $\sigma_{12} = \sigma_{32} = 0$, thus only the $i = 2$ term in Eq. (3.38) is non-zero. The stress ahead of the crack is from Eq. (2.49) $\sigma_{22}(x_1, 0) = \sigma_{\theta\theta}(x_1, 0) = \frac{K_I}{\sqrt{2\pi x_1}}$. The displacement is $u_2(\Delta a - x_1, \pi) = -u_\theta(\Delta a - x_1, \pi) = \frac{4K_I}{E'}\sqrt{\frac{r}{2\pi}}$. Substituting into Eq. (3.38), evaluating the integral and simplifying

$$G = \frac{K_I^2}{E'}. \tag{3.40}$$

Recall that $E' = E$ for plane stress and $E' = E/(1 - \nu^2)$ for plane strain.

3.6.3 Mode II Loading

Similarly, for Mode II loading,

$$G = \frac{K_{II}^2}{E'}. \tag{3.41}$$

3.6.4 General Loading (2D Crack)

For a 2D crack under general loading the stress and displacement fields are a superposition of the Mode-I, II, III fields and the energy release is

$$G = \frac{K_I^2}{E'} + \frac{K_{II}^2}{E'} + \frac{K_{III}^2}{2\mu}. \tag{3.42}$$

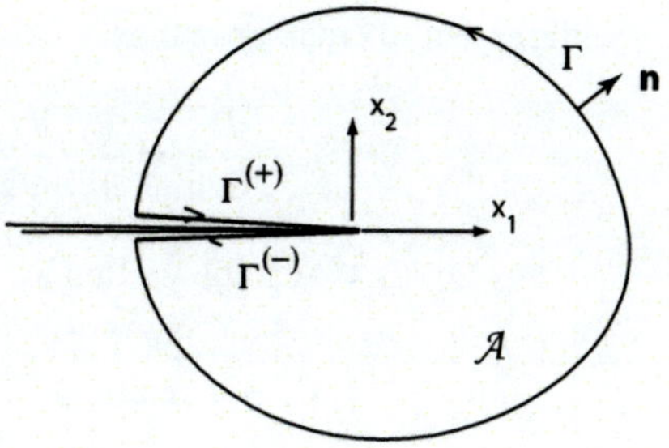

Fig. 3.7 To calculate the energy release rate in terms of the J integral take $\mathscr{A}$ to be any fixed region surrounding the crack tip

3.7 Contour Integral for G (J-Integral)

3.7.1 Two Dimensional Problems

An alternative to the crack closure integral is to take an integral on a contour that surrounds the crack tip. This integral will represent the energy that flows to the crack tip and will be shown to be equivalent to the crack closure integral. The original derivation is given by Rice [3]. Generalizations and further discussion are given in [4–6].

Begin by specializing Eq. (3.32) to 2D cracks that grow straight ahead, i.e. in the x_1 direction. In this case $\Delta\mathscr{V} \to 0$ and the volume integrals become area integrals, i.e. $\int_{\mathscr{V}}(\cdot)dV \to \int_{\mathscr{A}}(\cdot)\,dA$. The increase $\Delta\mathscr{S}$ in crack surface area will correspond to an increase in crack length of Δa. Thus,

$$G = -\frac{\partial \Pi}{\partial a} = \lim_{\Delta a \to 0} \frac{1}{\Delta a} \int_{\mathscr{A}} \big\{ (\sigma_{ij}^0 + \Delta\sigma_{ij})\Delta\gamma_{ij} - \big[W(\gamma^0 - \Delta\gamma) - W(\gamma^0)\big]\big\}\, dA. \tag{3.43}$$

The region $\mathscr{A}$ is the entire body. However since the integrand is second order in Δa, as $\Delta a \to 0$ it is sufficient to consider $\mathscr{A}$ as any finite region in which the crack is embedded.

This region will have a boundary $\Gamma + \Gamma^{(+)} + \Gamma^{(-)}$, where Γ is a contour that starts and stops on the crack line, and $\Gamma^{(+)}$ and $\Gamma^{(-)}$ lie along the top and bottom crack faces, as shown in Fig. 3.7.

$\Delta a \to 0$. As in the derivation of Eq. (3.34), by applying the divergence theorem, equilibrium and $\sigma_{ij} n_j = t_i$, and noting that $t_i^0 + \Delta t_i = 0$ on the crack faces,

$$\lim_{\Delta a \to 0} \frac{1}{\Delta a} \int_{\mathscr{A}} (\sigma_{ij}^0 + \Delta\sigma_{ij})\Delta\gamma_{ij}\, dA = \lim_{\Delta a \to 0} \frac{1}{\Delta a} \int_{\Gamma} (t_i^0 + \Delta t_i)\Delta u_i \, d\Gamma. \tag{3.44}$$

Consider a fixed coordinate system, (x_1, x_2) and coordinates (x_1', x_2') that move with the crack tip, $x_1' = x_1 - a$ and $x_2' = x_2$ for straight ahead crack growth. Any field quantity f can be written with respect to the crack tip as $f = f(x_1', x_2', a)$ or with respect to stationary coordinates as $f(x_1 - a, x_2, a)$. The total derivative of f

with respect to crack growth is

$$\frac{Df}{Da} = \frac{\partial f}{\partial x'_1}\frac{\partial x'_1}{\partial a} + \frac{\partial f}{\partial a} = -\frac{\partial f}{\partial x_1} + \frac{\partial f}{\partial a}.$$

Thus Eq. (3.44) can be written as

$$\lim_{\Delta a \to 0} \frac{1}{\Delta a} \int_{\Gamma} (t_i^0 + \Delta t_i)\Delta u_i \, d\Gamma = \int_{\Gamma} t_i^0 \frac{Du_i}{Da} \, d\Gamma + \int_{\Gamma} \lim_{\Delta a \to 0} \Delta t_i \frac{\Delta u_i}{\Delta a} \, d\Gamma.$$

The integrand of the second term is of order Δa and hence will go to zero as $\Delta a \to 0$, leaving

$$\lim_{\Delta a \to 0} \frac{1}{\Delta a} \int_{\Gamma} (t_i^0 + \Delta t_i)\Delta u_i \, d\Gamma = \int_{\Gamma} t_i^0 \left(-\frac{\partial u_i}{\partial x_1} + \frac{\partial u_i}{\partial a} \right) d\Gamma. \tag{3.45}$$

To simplify the second term of Eq. (3.43) note that

$$\lim_{\Delta a \to 0} \frac{1}{\Delta a} \int_{\mathscr{A}} [W(\gamma^0 + \Delta\gamma) - W(\gamma^0)] \, dA = \int_{\mathscr{A}} \frac{DW}{Da} \, dA.$$

Differentiating as described above,

$$\begin{aligned} \int_{\mathscr{A}} \frac{DW}{Da} \, dA &= \int_{\mathscr{R}} \left(\frac{\partial W}{\partial a} - \frac{\partial W}{\partial x_1} \right) dA = \int_{\mathscr{A}} \left(\frac{\partial W}{\partial \gamma_{ij}} \frac{\partial \gamma_{ij}}{\partial a} - \frac{\partial W}{\partial x_1} \right) dA \\ &= \int_{\mathscr{A}} \left(\sigma_{ij} \frac{\partial u_{i,j}}{\partial a} - \frac{\partial W}{\partial x_1} \right) dA = \oint_{\Gamma + \Gamma^{(+)} + \Gamma^{(-)}} \left(t_i \frac{\partial u_i}{\partial a} - W n_1 \right) d\Gamma, \end{aligned}$$

using the divergence theorem in the last line. Note that on $\Gamma^{(+)}$ and $\Gamma^{(-)}$ $n_1 = 0$ and $t_i = 0$, thus there is no contribution to the contour integral from these portions and hence

$$\lim_{\Delta a \to 0} \frac{1}{\Delta a} \int_{\mathscr{A}} \left[W(\gamma^0 + \Delta\gamma) - W(\gamma^0) \right] dA = \int_{\Gamma} \left(t_i \frac{\partial u_i}{\partial a} - W n_1 \right) d\Gamma. \tag{3.46}$$

Inserting Eqs. (3.44), (3.45) and (3.46) into Eq. (3.43) in the limit as $\Delta a \to 0$ we have the "J-integral" [3]:

$$G = \int_{\Gamma} \left(W n_1 - t_i \frac{\partial u_i}{\partial x_1} \right) d\Gamma = J. \tag{3.47}$$

The importance of the J-integral in fracture mechanics cannot be overstated. It is used in many contexts to compute energy flow to the crack tip, to estimate crack opening and is used as part of failure criteria for ductile materials. Much more of J will be seen in later chapters.

J has the following important properties: It is path independent, that is any path Γ that starts and ends on the crack faces will give the same value of J, see the exercises. The value of J does not depend on the direction of subsequent crack

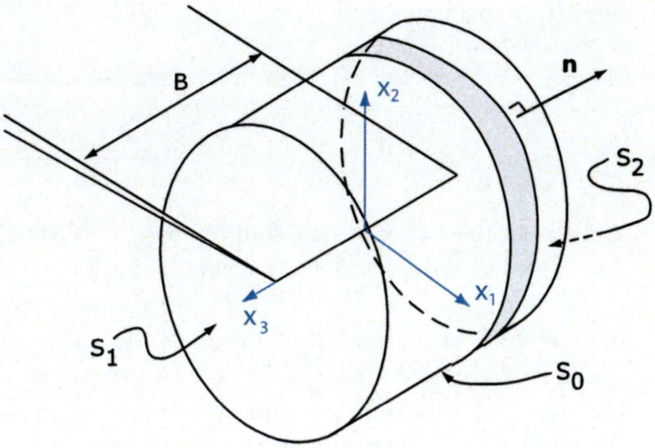

Fig. 3.8 Surface for computing J in 3D. S_1 and S_2 are on the free surfaces of a plate of thickness, b

growth, however, $J = G$ *only* for straight ahead, (x_1 direction) crack growth in elastic materials. (The energy release rate, Eq. (3.43) is given only for straight ahead crack growth.) Thus J has the same relation to the stress intensity factors as does G for straight ahead crack growth, namely for general loading,

$$J = \frac{K_I^2}{E'} + \frac{K_{II}^2}{E'} + \frac{K_{III}^2}{2\mu}. \tag{3.48}$$

3.7.2 Three-Dimensional Problems

The J integral can be extended to three dimensions. For example, consider a crack in a plate of thickness B as shown in Fig. 3.8. Surround the crack line with a surface $S = S_0 + S_1 + S_2$. Since $n_1 = 0$ and $t_i = 0$ on S_1 and S_2, the J-integral in this case will be

$$J = \frac{1}{B} \int_{S_0} (W n_1 - t_i u_{i,1})\, dA. \tag{3.49}$$

Note that in computational methods a local value of $J(x_3)$ can be defined by taking the 2D J for slices (for example shaded slice in the figure) along the crack line. The sum of these slices can be interpreted as the surface integral above.

3.7.3 Example Application of J-Integral

An experimental setup to study the fracture of elastomers consists of a long strip loaded in tension. To analyze this configuration let us approximate it as a thin, infinite strip of height h and thickness B containing a semi-infinite crack, as shown in Fig. 3.9. Assume that the boundary conditions are $u_1(x_1, \pm\frac{h}{2}) = 0$, $u_3(x_1, \pm\frac{h}{2}) = 0$

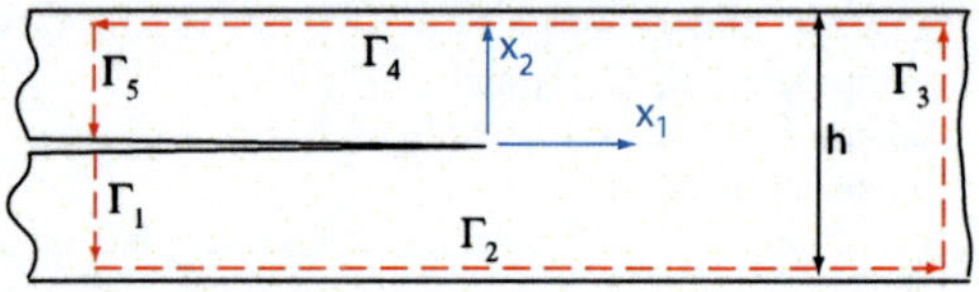

Fig. 3.9 Infinite strip with semi infinite crack

and $u_2(x_1, \pm\frac{h}{2}) = \pm\frac{\Delta u}{2}$. Assuming $B \ll h$ plane-stress can be assumed to hold away from the edges and away from the crack tip.

Choose $\Gamma = \Gamma_1 + \Gamma_2 + \Gamma_3 + \Gamma_4 + \Gamma_5$ to be along the path shown, i.e. with vertical sections far to the left and right of the crack and with horizontal sections just inside of the boundaries. On Γ_1 and Γ_5 there is no contribution to J since behind the crack the material is completely unloaded and hence $W = 0$ and $t_i = 0$ here. Along Γ_2 and Γ_4 there is also no contribution to J since $n_1 = 0$ and $u_{i,1} = 0$ since u_1 and u_2 are constants along the boundaries. Only Γ_3 contributes to J. Along Γ_3, $n_1 = 1$ and $t_i u_{i,1} = \sigma_{11} u_{1,1} + \sigma_{21} u_{2,1}$. Far to the right of the crack $\sigma_{12} = 0$ and $u_1 = 0$, thus $\int_{\Gamma_3} t_i u_{i,1}\, d\Gamma = 0$. This leaves only $J = \int_{\Gamma_3} W\, d\Gamma$. Away from the crack the material is in uniform, biaxial tension and hence $W =$ constant and $J = hW$. Assuming plane stress with $\gamma_{22} = \frac{\Delta u}{h}$ and $\gamma_{11} = 0$, the strain energy density is $W = \frac{1}{2}(\frac{\Delta u}{h})^2 \frac{E}{1-\nu^2}$, and hence

$$J = \frac{h}{2}\left(\frac{\Delta u}{h}\right)^2 \frac{E}{1-\nu^2}. \tag{3.50}$$

3.8 Exercises

1. Prove that $\sigma_{ij}\gamma_{ij} = \sigma_{ij} u_{i,j}$.
2. Derive Eq. (3.27).
3. A semi-circular sample of thickness $B = 20.7$ mm, radius $W = 50$ mm and elastic modulus, $E = 10^5$ N/mm^2 was analyzed using the finite element method. The geometry and mesh for the analysis are shown in Fig. 3.10. Applying a load of $P = 100$ N, the analysis was repeated for a number of crack lengths. The load-point displacement q vs. the crack length is tabulated in Table 3.1. Define a non-dimensional energy release rate by $\tilde{G} = \frac{GEB^2W}{P^2}$. Using the compliance method and taking numerical derivatives of the data in the table, tabulate and plot the non-dimensional energy release rate, $\tilde{G}$, vs. the non-dimensional crack length, a/W.
4. Prove that J is path independent. To do this, first show that the integral

$$\oint_\Gamma \left(W n_i - t_j \frac{\partial u_j}{\partial x_i} \right) d\Gamma = 0.$$

for any closed curve Γ that surrounds a simply connected region (i.e. no holes inside of Γ). Then, specialize the above to $i = 1$ and consider the Γ shown in Fig. 3.11, breaking up Γ into the four sectors suggested by the figure.

Table 3.1 Computed load point displacement vs. crack length for test sample shown in Fig. 3.10 with load $P = 100$ N

a, mm	q, mm
14	6.640×10^{-4}
18	8.787×10^{-4}
22	12.12×10^{-4}
26	17.37×10^{-4}
30	29.75×10^{-4}
34	43.85×10^{-4}
38	82.49×10^{-4}

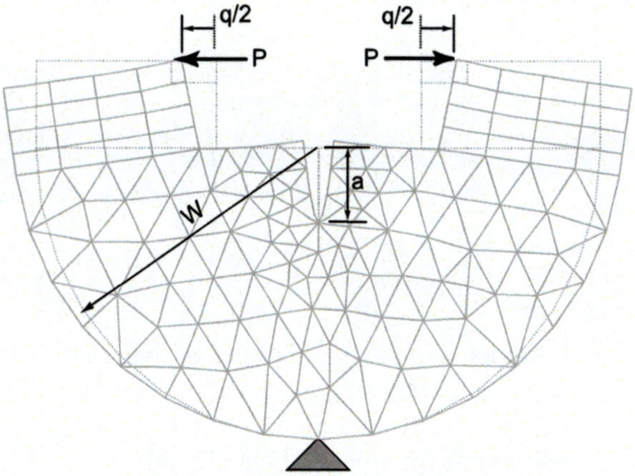

Fig. 3.10 FEM mesh for semi-circular test specimen. Outline of sample is shown in *dashed lines*. Deformed shape and mesh shown in *solid lines*. Displacement is magnified greatly

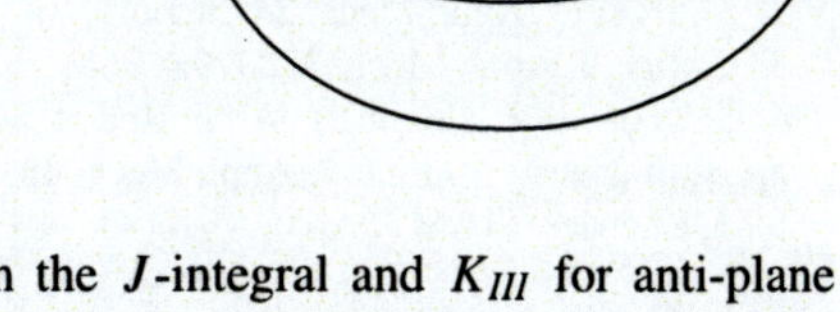

Fig. 3.11 To prove that J is path independent let $\Gamma = \Gamma_1 - \Gamma_2 + \Gamma^{(+)} + \Gamma^{(-)}$

5. Directly calculate the relation between the J-integral and K_{III} for anti-plane shear. You can, for example, choose a circular path around the crack tip and then simply substitute the asymptotic anti-plane shear fields into J. *Extension*: Instead of using the asymptotic fields, use the general expression (either the series expansion, Eq. (2.23) or the full-field solution, Eq. (2.33)) and show that the addition of the higher order terms does not change the value of J.
6. Calculate the J integral for the double cantilever beam specimen shown in Fig. 3.4.

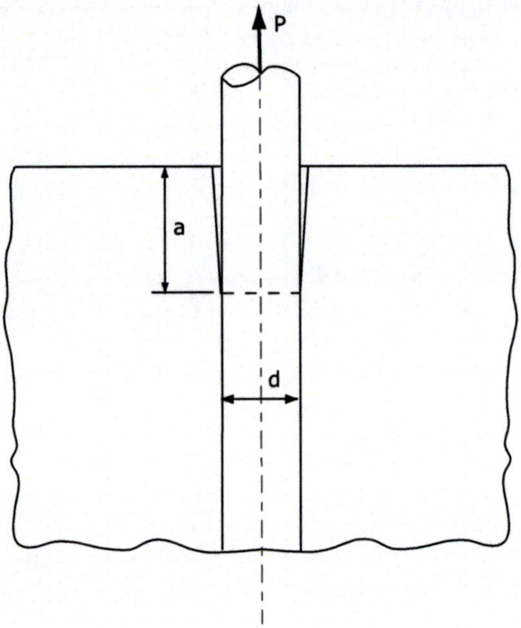

Fig. 3.12 Pull out of elastic fiber from rigid half-space

7. Consider a single, elastic fiber of diameter d, embedded in a rigid half-space. Suppose that the fiber is debonded from the substrate to a depth of a, as shown in Fig. 3.12. If a tensile force of P is applied to the fiber, calculate the energy release rate for growth of the debond.
8. Calculate J for an anti-plane shear crack with finite opening angle β. What does this result tell you?

References

1. J.R. Rice, in *Fracture an Advanced Treatise*, vol. 2, ed. by H. Liebowitz (Academic Press, New York, 1968), pp. 191–311, chap. 3
2. G. Irwin, J. Appl. Mech. **24**, 361 (1957)
3. J. Rice, J. Appl. Mech. **35**, 379 (1968)
4. J. Knowles, E. Sternberg, Arch. Ration. Mech. **44**, 187 (1972)
5. B. Budiansky, J. Rice, J. Appl. Mech. **40**, 201 (1973)
6. J. Knowles, SIAM J. Appl. Math. **41**, 401 (1981)

Chapter 4
Criteria for Elastic Fracture

Abstract Criteria for the growth of a pre-existing crack as generally based on reaching critical values of the stress intensity factors or energy release rate. Starting with the simplest case, that of tensile fracture in an elastic material, criteria for fracture under a range of conditions are explored. These include the initiation and evolution of cracks under mixed-mode loading, considerations of the stability of fracture, effects of temperature, fatigue crack growth and stress corrosion cracking.

4.1 Introduction

Elastic fracture should in the very narrowest sense mean that the only change to a material during fracture is atomic separation along the fracture surface. Everywhere else the material is unchanged, i.e. there is no re-arrangement of the structure of the material. Such cases are very rare, or perhaps even non-existent. A pragmatic definition of elastic fracture would be that the size of the "process zone", or region around the crack tip in which inelastic deformation takes place, be it plastic flow, micro-fracture etc., is very small compared to the dimensions of the test specimen of component being considered. In such cases the load-deflection relation of the component in the absence of crack growth is linear and linear elastic stress analysis is sufficiently accurate to describe the stress, strain and displacement fields. The discussion begins with consideration of Mode-I loading, followed by Mode-II and -III and mixed mode loadings.

4.2 Initiation Under Mode-I Loading

We assume from the start that we have a body with a sharp crack subjected to tensile loads. In the last chapter we showed that when a crack grows, energy is "released" from the body to the crack tip at a rate of G [J/m^2]. If we assume that a given amount of energy per unit area is required to grow a crack, then the resulting fracture criterion is that a crack will grow when the available energy release rate G is greater than or equal to the required energy, G_c, or

$$G \geq G_c. \tag{4.1}$$

A.T. Zehnder, *Fracture Mechanics*,
Lecture Notes in Applied and Computational Mechanics 62,
DOI 10.1007/978-94-007-2595-9_4,

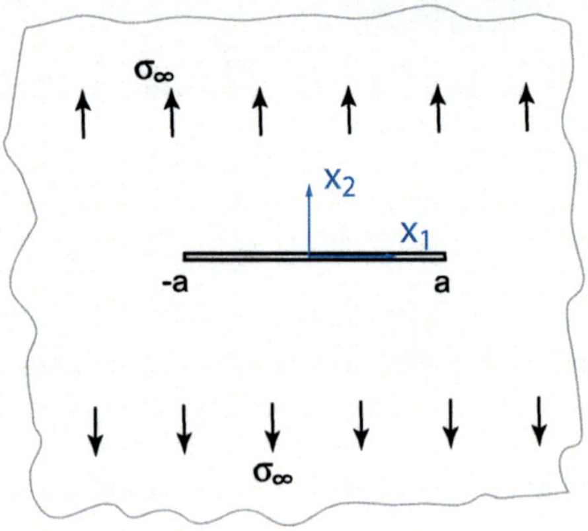

Fig. 4.1 Through crack of length $2a$ in a plate under tension. A uniform stress field $\sigma_{22} = \sigma_\infty$ is assumed to exist away from the crack

Note that $G = G(P, a, \text{geometry})$, i.e. the available energy release rate depends on the applied load, on the crack length (or area) and on the geometry of the body. The required energy, G_c is considered to be a material property and is called the "fracture energy" or "fracture toughness." Physically this energy combines the energy of the newly created surfaces with the energy dissipated in the process zone through plastic deformation, micro-cracking, friction, craze formation, void growth and so on. In anisotropic materials G_c may depend on the direction of crack growth. Consider fracture in wood for example, where cracks will prefer to grow along the grain rather than across it, demonstrating that the toughness along the grain is much less than across the grain. The above can be considered as a generalization of Griffith's fracture criterion in which G_c is equal to the energy of the newly created fracture surfaces, [1].

The canonical problem of the Griffith theory is the through crack in a plate under tension, as shown in Fig. 4.1. Assuming plane stress, G for this problem is [2]

$$G = \frac{\sigma_\infty^2 \pi a}{E}, \tag{4.2}$$

where σ_∞ is the tensile stress, E is the Young's modulus and a is the half crack length. Setting $G = G_c$ and solving, the predicted stress at the onset of fracture is

$$\sigma_f = \sqrt{\frac{G_c E}{\pi a}}. \tag{4.3}$$

To test the theory, Griffith had glass taken from test tubes blown into thin-walled spheres and cylinders. A glass cutter was used to introduce through cracks in the test samples, which were then annealed to eliminate any residual stresses due to cutting. The samples were then pressurized and the pressure at the point of unstable fracture recorded. The stress σ_f at fracture was then calculated for each test. The results, shown in Fig. 4.2, demonstrate that the failure stress is linear with $1/\sqrt{a}$ as predicted by the theory. The data in Fig. 4.2 are best fit with $G_c = 3.2$ J/m^2, corresponding to a critical stress intensity factor of $K_C = \sqrt{EG_c} = 0.45$ MPa$\sqrt{\text{m}}$.

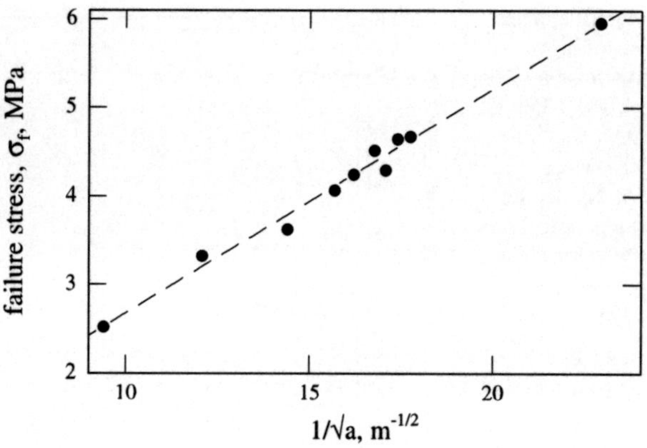

Fig. 4.2 Fracture stress vs. $1/\sqrt{a}$ from Griffith's experiments on cracked glass tubes and spheres [1]. a is the half crack length, and σ_f is the measured failure stress. Data are shown as *points*. The *line* is a linear fit to the data. Note that the fracture strength will not increase indefinitely as a gets smaller. The strength of a "perfect" structure is limited by the intrinsic bond strength of the material. In the case of glass the theoretical maximum strength is on the order of 1 GPa, thus Griffith's measured fracture strengths are about an order of 100× lower than the theoretical strength. The sensitivity of glass and other brittle materials to small flaws is what ultimately limits their practical strengths. Thin fibers, which due to their very small dimensions cannot contain large cracks, will generally exhibit improved tensile strengths relative to bulk glass

Table 4.1 Selected toughness values for nominally brittle materials

Material	K_{IC}, MPa$\sqrt{\text{m}}$	G_c, J/m^2
Borosilicate glass	0.8	9.
Alumina 99% polycrystalline	4.0	39
Zirconia-toughened alumina	6.	90.
Yttria partially stabilized zirconia	13	730
Aluminum 7075-T6	25	7800.
AlSiC metal matrix composite	10.	400.
Epoxy	0.4	200

For now only materials with *isotropic fracture and elastic properties* will be considered. Such materials have no preferred direction of crack growth. As will be shown later, when subjected to a tensile load a crack (at least in 2D) will grow straight ahead and hence we can make the connection between K_I and G, Eq. (3.40). Since in this case K_I and G are equivalent the fracture criterion can also be presented as

$$K_I \geq K_C \tag{4.4}$$

where the "fracture toughness", $K_C = \sqrt{E'G}$. A selection of typical fracture toughness values (for nominally brittle materials) is given in Table 4.1.

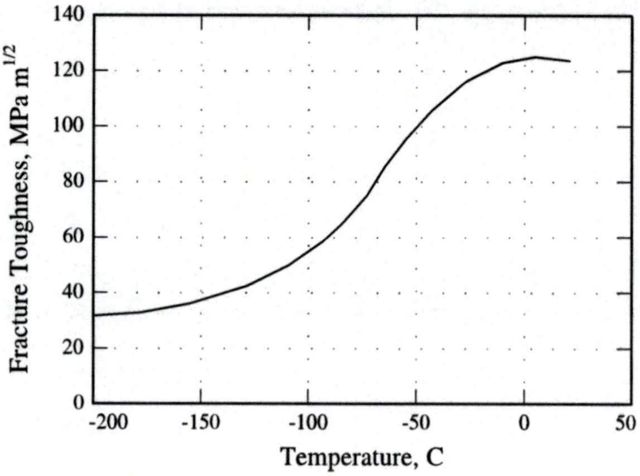

Fig. 4.3 Fracture toughness vs. temperature for high strength steel with yield strength 1175 MPa. 0.35% C, 0.65% Mn, 0.35% Si, 0.80% Cr, 0.30% Mo, 0.10% V, 1.26% Ni, bal. Fe. Note that the toughness drops off dramatically below approximately −40°C. Adapted from [3]

Environmental conditions play an important role in fracture and most other aspects of material behavior. For example the fracture toughness of metals and polymers is generally reduced as the temperature is reduced. In some cases this effect can be drastic. For example, in Fig. 4.3 the toughness of a high strength steel alloy is plotted vs. temperature. Above 0°C the material is ductile and has a high fracture toughness of over 120 MPa$\sqrt{\text{m}}$. However, as the temperature drops below −40°C the fracture becomes more brittle and toughness drops to less than 40 MPa$\sqrt{\text{m}}$. Since energy is the square of K_I this implies that the energy needed for fracture drops by almost a factor of 10 at low temperatures. Avoiding brittle fracture at low temperatures is of key importance in modern structural design, be it for railcars traversing North Dakota in winter or ships crossing the North Atlantic.

Ductile and brittle fractures are characterized not only by low and high toughness values, but also by the physical mechanisms of fracture. Ductile fracture in metals generally occurs through a process of void formation,void growth under tensile stress followed by coalescence of the voids leading to macroscopic fracture. A typical fracture surface, see Fig. 4.4 exhibits a dimpled appearance. Brittle fracture in metals can be transgranular, i.e. cutting across grains or intergranular, where the crack grows around the grains, or a mix of both. In either case the fracture surface is generally smoother. An example of intergranular growth is shown in Fig. 4.5.

The simple fracture criteria above will let you determine if a crack will grow or not, but it does not tell us anything about how fast, how far, or in what direction the crack will grow. These topics are addressed in subsequent chapters and sections.

4.3 Crack Growth Stability and Resistance Curve

If, as a crack extends from a pre-existing flaw, the available G falls below the toughness, G_C, then the crack will stop growing (arrest) and will continue to grow only if the loading or other conditions change. If as the crack grows, G increases and becomes ever higher than G_C the crack will become unstable, rapidly growing until

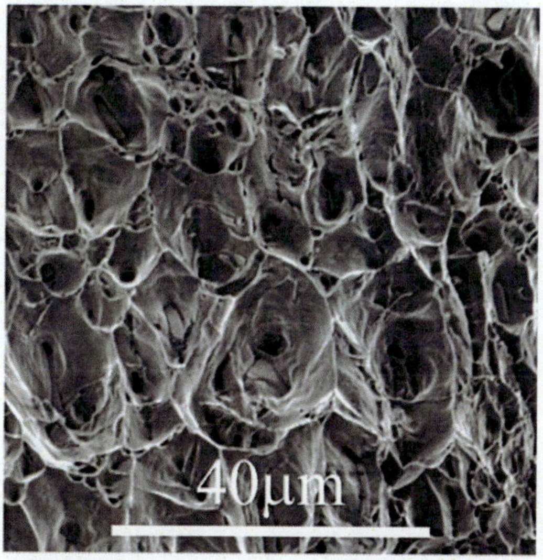

Fig. 4.4 Ductile fracture surface, aluminum 6061. The dimpled surface is a characteristic of ductile fracture in which failure proceeds through a process of void formation, void growth under tensile stresses followed by coalescence of voids leading to fracture. ©DoITPoMS Micrograph Library, University of Cambridge, http://www.doitpoms.ac.uk, micrograph 138, K Clark, University of Cambridge, used by permission

Fig. 4.5 HY100 forging steel, notched impact specimen, failed through brittle intergranular fracture. ©DoITPoMS Micrograph Library, University of Cambridge, http://www.doitpoms.ac.uk, micrograph 146, S Munday, University of Cambridge, used by permission

the body is completely fractured. The prediction, or better yet, prevention of unstable crack growth (except when it is desirable as in some manufacturing operations) is of primary concern in mechanical and structural design and hence warrants careful attention.

The stability of crack growth depends on both the characteristics of the material and on the geometry and nature of the loading. For example, from the analysis of the DCB specimen it is easily shown that if the applied load is fixed then G grows as a^2, Eq. (3.26), whereas if the displacement is fixed G decreases as $1/a^4$, Eq. (3.27).

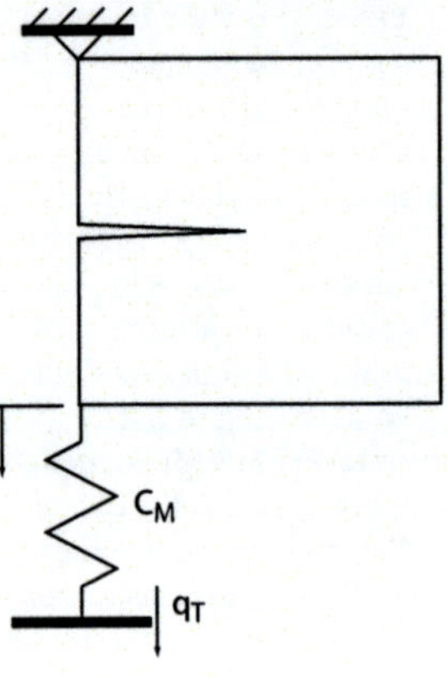

Fig. 4.6 General loading by a compliant system

Thus it is far more likely for fracture to be stable under conditions of fixed displacement loading than under conditions of fixed force.

4.3.1 Loading by Compliant System

In most applications the component is not under fixed force or fixed displacement loadings, but rather under an intermediate state in which as the crack propagates and the component becomes more compliant, load is transferred to other parts of a system. In terms of generalized forces and displacements such a general loading can be represented as a body loaded by a spring with compliance C_M, to which a fixed displacement, q_T, is applied, as sketched in Fig. 4.6. The displacement of the body is q and the load is

$$P = \frac{q_T}{C + C_M},$$

where C is the compliance of the body. Loading of the body by fixed displacement is equivalent to $C_M = 0$, while loading by fixed force is equivalent to $C_M = \infty$.

The energy release rate is unchanged from the previous results, Eq. (3.25) i.e. $G = \frac{1}{2}P^2\frac{\partial C}{\partial s}$. However, the rate of change of G *is* affected. Substituting $P = \frac{q_T}{C+C_M}$ into Eq. (3.25) and differentiating

$$\frac{\partial G}{\partial s} = -P^2\frac{(\partial C/\partial s)^2}{C + C_M} + \frac{1}{2}P^2\frac{\partial^2 C}{\partial s^2}. \tag{4.5}$$

Since the first term in Eq. (4.5) is always < 0, $\partial G/\partial s$ is always smaller for the case of finite C_M than for dead weight loading $C_M = \infty$. As the "machine" becomes stiffer (C_M decreases) the first term becomes larger and eventually, $\partial G/\partial s < 0$, stabilizing crack growth.

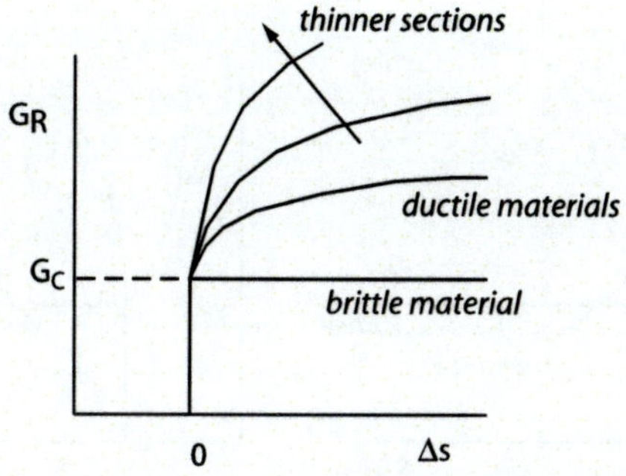

Fig. 4.7 Schematic of resistance curve (R-curve), $G_R(\Delta s)$. Generally G_R increases with crack extension. It may or may not reach a steady state. For elastic-plastic materials, G_R increases much more for plane stress than for plane strain problems, due to the decreased constraint and hence greater ease of plastic flow in plane stress. Note that in 2D, $\Delta s = B\Delta a$, where B is the thickness of the plate

4.3.2 Resistance Curve

In some cases crack growth can be stable even when $\partial G/\partial s > 0$. How so? In some cases, a material's resistance to fracture increases with crack extension. In 2D, let $G_R(\Delta a)$ be the energy release rate required to propagate a crack. This function, called the "resistance curve" is sketched in Fig. 4.7. The physical sources of an increasing G_R curve are numerous; two examples are given here: (1) For elastic-plastic materials, the level of crack tip strain for the same applied stress intensity factor, is less for propagating cracks that for stationary cracks. If a specific strain ahead of the crack is needed to grow the crack, this implies that the applied stress intensity must be increased for the growing crack. (2) For fiber or particulate reinforced composites, as the crack grows there may be fibers in the wake of the crack tip that bridge across the crack tip, applying a closing force near the crack tip. To overcome these closing forces the externally applied stress intensity factor must be increased.

For perfectly brittle materials, G_R is a constant. For other materials, G_R may rise and then reach a steady state value or it may continue to increase. In ductile metals it is generally found that the resistance curve increases at a faster rate for tests done on thin sheets than for tests done on thick sections. In small scale yielding, K_I and G are related by Eq. (3.40) and thus the resistance curve could also be presented in terms of $K_R(\Delta s)$.

Regardless of the source of the resistance curve, a criterion for crack stability can be stated. In 2D, for the crack to continue to advance,

$$G(s) = G_R(s - s_0), \tag{4.6}$$

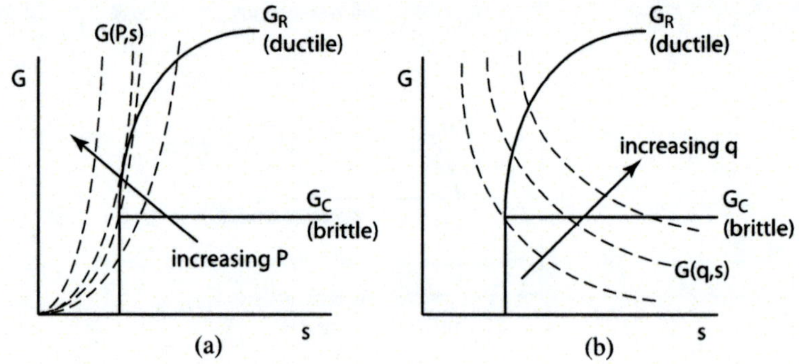

Fig. 4.8 Available energy release rate for (**a**) fixed force loading and (**b**) fixed displacement loading superimposed with crack growth resistance, G_R. Loading with fixed force will be unstable after a small amount of crack growth even for a material with a rising R-curve. Crack growth is always stable under fixed displacement loading. Most applications would involve loading through a compliant system, thus the stability will be between the two extremes above

where $s_0 = B \cdot a_0$, a_0 is the initial crack length, $s = B \cdot a$, and a is the current crack length. The condition for stability of crack growth is

$$\frac{\partial G}{\partial s} < \frac{\partial G_R}{\partial \Delta s}. \tag{4.7}$$

Unstable crack growth may occur when

$$\frac{\partial G}{\partial s} \geq \frac{\partial G_R}{\partial \Delta s}. \tag{4.8}$$

Consider the two extreme cases of fixed load and fixed displacement loading. In the case of fixed load, G is an increasing function of a, e.g. Eq. (3.26) for the DCB sample. For fixed displacement G decreases with a, e.g. Eq. (3.27) for the DCB sample. In all cases G increases as P^2 or q^2. The available energy release rates $G(P,a)$ for the fixed load case and $G(q,a)$ for the fixed displacement case are superimposed with the G_R resistance curves in Figs. 4.8(a,b).

In the case of *fixed displacement* loading, the crack growth is always stable, i.e. the crack will begin to grow when the applied displacement is large enough that $G(q,a_0) = G_R(0) \equiv G_C$. To continue to grow the crack the applied displacement must be increased to satisfy Eq. (4.6).

In the case of *fixed force* loading, crack growth will be unstable for brittle materials, since once the load has increased to a value such that $G(P,a) = G_C$, any increment in crack growth will increase G to a value above G_C. However, if the material has a rising resistance curve, then it may be possible for the crack to initiate at $G = G_C$ and then grow stably for a small distance before the G curve is tangent to the G_R curve. Note that once the fracture becomes unstable, the crack can grow very fast; at speeds that approach the sound speed of the material.

In many real applications the loading is somewhere between fixed load and fixed displacement, thus G may increase or decrease with respect to a and the stability will depend on the relative stiffness of the loading and on the slope of the R curve. Thus, although for a brittle material stable crack growth will only be possible under fixed displacement loading where $\frac{\partial G}{\partial s} < 0$, for a material with rising R-curve, stable crack growth is possible if

$$\frac{\partial G}{\partial s} = -P^2 \frac{(\partial C/\partial s)^2}{C + C_M} + \frac{1}{2} P^2 \frac{\partial^2 C}{\partial s^2} < \frac{\partial G_R}{\partial \Delta s}.$$

4.4 Mixed-Mode Fracture Initiation and Growth

Cracks in components whose directions of principal stress vary spatially or in components made of a material with an anisotropic fracture toughness (toughness depends on direction of fracture plane) will generally propagate along a curving surface as the crack seeks out its path. The prediction not only of when and how far a crack will grow, but of its path is important in the analysis of potential failures. For example, if a gear tooth was cracked would the crack propagate across the tooth, just breaking it off? Or would the crack propagate into the hub of the gear, causing the entire gear to fly apart? Would a crack in the skin of a pressurized aircraft fuselage grow straight and unzip the entire fuselage or would it curve and be contained with one bay (a section of the fuselage bounded by the circumferential frames and the longitudinal stringers)?

Such problems involve complex, 3D geometries. The cracks in such cases generally have a mix of Mode-I,-II,-III loadings that varies along the crack front and that may vary as well during a cycle of loading. As a start to understanding such problems we will start with 2D mixed-mode loadings.

If as sketched in Fig. 4.9, a crack is subjected to a combination of Mode-I and Mode-II loadings (mixed-mode loading) the crack will generally not propagate straight ahead. The exception would be if the line along $\theta = 0$ is a very weak plane of the material, for example wood grain, or a weak bond line.

Far away from the crack the stress is $\sigma_{22} = \sigma_a \sin^2 \beta$, $\sigma_{12} = \sigma_a \sin\beta \cos\beta$, $\sigma_{11} = \sigma_a \cos^2 \beta$. The resulting stress intensity factors are $K_I = \sigma_a \sqrt{\pi a} \sin^2 \beta$, $K_{II} = \sigma_a \sqrt{\pi a} \sin\beta \cos\beta$. At what stress level, σ_a will the crack begin to grow and in what direction?

Theories for mixed-mode fracture include: (1) maximum circumferential stress [4], (2) minimum strain energy density [5], (3) maximum energy release rate [6] and (4) local symmetry [7]. Only the theory of maximum circumferential stress and of maximum energy release rate will be discussed here.

4.4.1 *Maximum Hoop Stress Theory*

The maximum circumferential, or hoop, stress theory postulates that a crack will grow in the direction, θ^*, of maximum hoop stress, $\sigma_{\theta\theta}$, when $\sqrt{r}\sigma_{\theta\theta}(r, \theta^*) \geq$ Const. Assuming that the constant is the same for mixed-mode loading as for pure

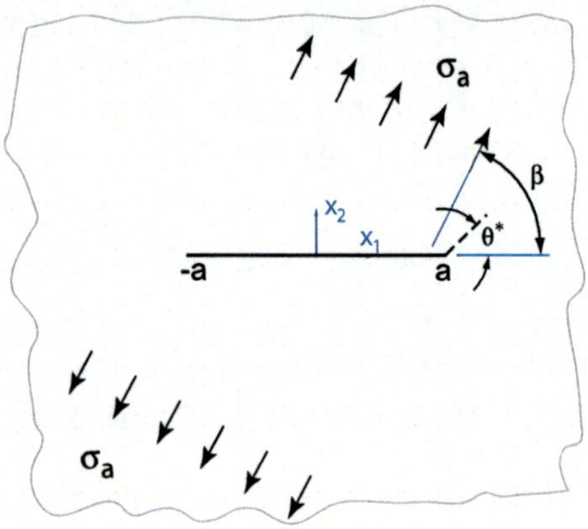

Fig. 4.9 2D crack under mixed-mode loading. $K_I = \sigma_a\sqrt{\pi a}\sin^2\beta$, $K_{II} = \sigma_a\sqrt{\pi a}\sin\beta\cos\beta$. Direction of next step of crack growth is shown as *dashed line on right* with direction θ^* from the crack line

Mode-I loading, from the Mode-I criterion $K_I \geq K_{IC}$, the mixed-mode criterion can be written as

$$\sqrt{r}\sigma_{\theta\theta}(r,\theta^*) \geq \frac{K_{IC}}{\sqrt{2\pi}}. \tag{4.9}$$

The directional criterion is that the crack will grow in the direction θ^* that satisfies

$$\frac{\partial\sigma_{\theta\theta}}{\partial\theta} = 0, \qquad \frac{\partial^2\sigma_{\theta\theta}}{\partial\theta^2} < 0. \tag{4.10}$$

Combining Eqs. (2.49) and (2.61) and re-arranging, the hoop stress can be written as

$$\sigma_{\theta\theta}(r,\theta) = \frac{1}{\sqrt{2\pi r}}\left(K_I\cos\frac{\theta}{2}\frac{1+\cos\theta}{2} - K_{II}\frac{3}{2}\sin\theta\cos\frac{\theta}{2}\right). \tag{4.11}$$

Substituting Eq. (4.11) into Eqs. (4.9) and (4.10) the direction of crack growth and the envelope of failure, expressed in terms of $(K_I/K_{IC}, K_{II}/K_{IC})$ can be determined. The results are plotted in Fig. 4.10.

Comparison of this theory with experimental results shows that the maximum hoop stress theory predicts the angle of crack growth well but somewhat underestimates the envelope of failure. Nonetheless, at least for crack growth angle the maximum hoop stress theory is quite accurate and is easily implemented in fracture simulations.

In elementary mechanics of materials one learns that brittle fracture will occur along the plane of maximum tensile stress. This criterion is in fact the same as the maximum hoop stress theory. Why then, is the relation between θ and β not a straight line? Or stated another way, why under pure Mode-II loading, where the direction of maximum principal stress is 45 deg to the crack line, does the crack grow at $\theta = -70.6$ deg and not -45 deg? The answer is that the presence of the crack disturbs the stress field, hence changing the directions of maximum principal stress. It is interesting to note however, that the path of the crack will evolve to lie along the plane of maximum principal stress. To demonstrate this a numerical

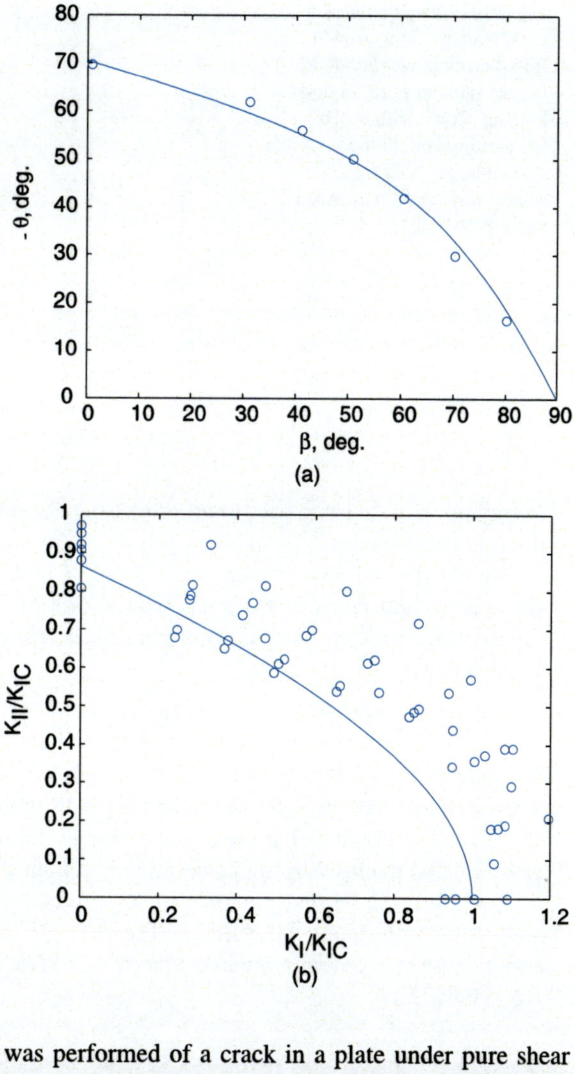

Fig. 4.10 (**a**) Angle of crack propagation predicted by maximum hoop stress theory versus load application angle. (**b**) Failure envelope predicted by maximum hoop stress theory. The theory predicts that under pure Mode-II loading the crack will grow at an angle of $\theta = -70.6$ deg at a stress intensity factor of $K_{II} = 0.87 K_{IC}$. Experimental data for kink angles and mixed-mode fracture toughness are taken from [4]. Theory predicts kink angles well and provides a lower bound to the mixed-mode fracture toughness envelope

(finite element) calculation was performed of a crack in a plate under pure shear (pure Mode-II) loading. Using the maximum hoop stress criterion the tip of the crack is moved ahead in small increments. The result, shown in Fig. 4.11 shows that the crack initially kinks at -70.6 deg, but then gently turns and grows at -45 deg.

4.4.2 Maximum Energy Release Rate Criterion

The maximum energy release rate criterion states that the crack will propagate so as to maximize the energy release. Since $G = -\frac{\partial \Pi}{\partial s}$, this criterion is equivalent to saying that the crack grows so as to minimize the potential energy of the body, corresponding to the thermodynamic idea that equilibrium systems seek their local

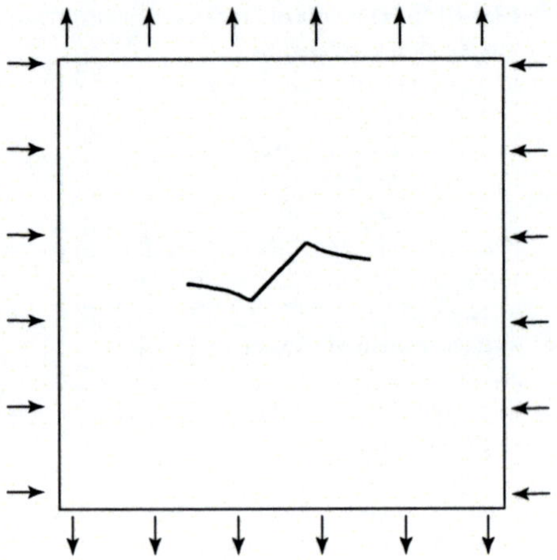

Fig. 4.11 Computational simulation of crack growth. Initial crack (*straight line* in figure) is under pure Mode-II loading. Crack path evolves to be orthogonal to the maximum far-field tensile stress. Simulation performed using FRANC [8]

energy minima. In principal this criterion can be applied in 3D and to crack surfaces of arbitrary shape.

Wu [6] studied this problem for 2D cracks under Mode-I,-II loading. He considers a straight crack that extends with a kink of length ϵ and angle θ from the pre-existing crack tip. The criterion can be stated as the crack will kink at the angle θ^* such that

$$\frac{\partial G(\theta^*)}{\partial \theta} = 0, \qquad \frac{\partial^2 G(\theta^*)}{\partial \theta^2} < 0, \quad \text{and } G(\theta^*) \geq G_C \tag{4.12}$$

where $G(\theta) \equiv \lim_{\epsilon \to 0} \frac{1}{\epsilon}(\Pi_Z - \Pi)$, and Π_Z is the potential energy for the kinked (or "z-shaped") crack and Π is the potential energy for the original crack. $G(\theta)$ cannot be calculated in closed-form, however an excellent approximation can be obtained.

The results show that the kink angle predicted by the energy release rate and hoop stress criteria are quite similar. For example in pure Mode-II loading the energy release rate criterion predicts that $\theta^* = -75.6°$ and that fracture occurs when $K_{II} \geq 0.817 K_{IC}$.

4.4.3 Crack Path Stability Under Pure Mode-I Loading

All of the above criteria predict that a crack under pure Mode-I loading will continue to propagate straight ahead. However what would happen if, for example a crack were to grow under a far field stress, with respect to the coordinate system in Fig. 4.9, of $\sigma_{11} = 2$, $\sigma_{22} = 1, \sigma_{12} = 0$. In this case $K_I = \sqrt{\pi a}$ and $K_{II} = 0$ and the above theories predict that the crack will grow straight ahead, i.e. in the x_1 direction. However, the maximum tensile stress away from the crack is not in the x_1 direction, but in the x_2 direction. So perhaps the crack would rather grow in the x_2 direction so that the material fractures on the plane of maximum tensile stress.

The above brings up the question of crack-path stability. This problem can be analyzed by considering a semi-infinite crack in a 2D stress field. If there is a kink

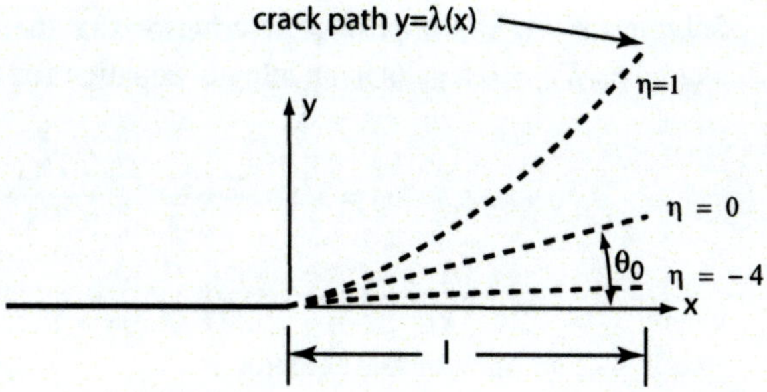

Fig. 4.12 Path of crack growth and dependence on T stress parallel to the main crack ($\eta = 2\sqrt{2}T/k_I$, $\theta_0 = 0.2$)

at the crack tip, will the crack return to its original path or will the crack turn away further from its original path?

Cotterell and Rice [9] developed a first order method to calculate the stress intensity factors at the tip of a slightly curved or kinked crack and applied it to predict the stability of crack paths. For a semi-infinite crack with an extension of length l and path $y = \lambda(x)$, see Fig. 4.12, they show that the stress intensity factors for the extended crack are

$$\begin{Bmatrix} K_I \\ K_{II} \end{Bmatrix} = \left(\frac{2}{\pi}\right)^{1/2} \int_{-\infty}^{l} \left[\begin{Bmatrix} t_y - \lambda'(l)t_x \\ t_x + \lambda'(l)t_y \end{Bmatrix} \frac{1}{(l-x)^{1/2}} + \begin{Bmatrix} t_x \\ t_y \end{Bmatrix} \frac{[\lambda(l) - \lambda(x) - \lambda'(l)(l-x)]}{2(l-x)^{3/2}} \right] dx, \tag{4.13}$$

where an (x, y) coordinate system is placed at the tip of the pre-existing crack, Fig. 4.12a, and t_x and t_y are the tractions on the new crack line necessary to remove the stresses that exist prior to crack extension, i.e. when the crack tip is at $(x, y) = (0, 0)$. From Eqs. (2.49) and (2.61) (including the constant stress term) the stress field on the x-axis is

$$\begin{aligned} \sigma_{yy}(x, 0) &= \frac{k_I}{\sqrt{2\pi x}}, \\ \sigma_{xx}(x, 0) &= \frac{k_I}{\sqrt{2\pi x}} + T, \\ \sigma_{xy}(x, 0) &= \frac{k_{II}}{\sqrt{2\pi x}}, \end{aligned} \tag{4.14}$$

where $x > 0$, k_I, k_{II} are the stress intensity factors for the original crack tip and T is the constant stress term in Eq. (2.49) ($T = 4A_0$). To first order in λ, the tractions are

$$\begin{aligned} t_y &= \frac{1}{\sqrt{2\pi x}} \left[k_I + \frac{\lambda(x)}{2x} k_{II} - \lambda'(x) k_{II} \right], \\ t_x &= \frac{1}{\sqrt{2\pi x}} \left[k_{II} + \frac{\lambda(x)}{2x} k_I - \lambda'(x) k_I \right] - \lambda'(x) T. \end{aligned} \tag{4.15}$$

Solving for K_{II} and imposing the criterion that the crack propagates along a path that holds $K_{II} = 0$ results in an integral equation for the path $\lambda(x)$

$$\theta_0 = \lambda'(l) - \frac{\beta}{\sqrt{\pi}} \int_0^l \frac{\lambda'(x)}{(l-x)^{1/2}} dx, \tag{4.16}$$

where $\theta_0 = -2k_{II}/k_I$ and $\eta = 2\sqrt{2}T/k_I$ is a normalized T stress with units of $1/\sqrt{L}$. The solution of this equation is

$$\lambda(x) = \frac{\theta_0}{\eta^2}\left[\exp(\eta^2 x)\,\mathrm{erfc}(-\eta x^{1/2}) - 1 - 2\eta\left(\frac{x}{\pi}\right)^{1/2}\right], \tag{4.17}$$

One can interpret θ_0 as the initial angle of crack growth.

As an example, the path for a crack with $\theta_0 = 0.2$ is shown in Fig. 4.12 for $\eta = \{-4,\ 0,\ 1\}$. In the case $\eta = -4$ ($T < 0$), the crack turns back towards its initial path. For $\eta = 1$, ($T > 0$), the crack turns away from its original path. For $\eta = 0$, ($T = 0$), the crack continues on a straight the line in the direction of its perturbation.

The stability of a straight crack under Mode-I loading can be deduced from the above analysis. Suppose that the crack has a slight kink or that the loads are not perfectly aligned orthogonal to the crack. In this case k_{II} will differ slightly from zero and hence $\theta_0 \neq 0$. If $T > 0$ the path will deviate, from straight. However, if $T < 0$, the crack will turn back. In addition, the rate of deviation increases strongly with T/k_1. Experimental results on crack paths under biaxial tension [10] are in agreement with this prediction.

As an additional illustration of the effect of T stress on crack path stability consider an initially tilted crack under biaxial tension, Fig. 4.13. The crack path is computed using a series of finite element simulations in which the crack is incremented by a small amount according to the maximum hoop stress theory [8]. With negative T stress the crack evolves to a horizontal path. When $T = 0$ the crack path is neutral, i.e. it continues along the original path. For $T > 0$ the crack path diverges and eventually will evolve to a vertical path, orthogonal to the direction of maximum tensile stress. Note that compression parallel to the crack stabilizes the crack path. This result can be exploited in experiments where one may wish to constrain the crack path.

A visually compelling illustration of crack path stability is the transition of cracks in glass under thermal stresses from straight to sinusoidal paths [11, 12]. Glass slides with a small prenotch were heated and then gradually immersed from one end into a water bath. As the temperature difference and speed at which the slides were immersed in the water were varied, the crack pattern transitions from straight to curved as shown in Fig. 4.14. Although the overall stress field is quite complex, a rough explanation for the oscillating crack is biaxial tension created by cooling as the slide is immersed in the water bath.

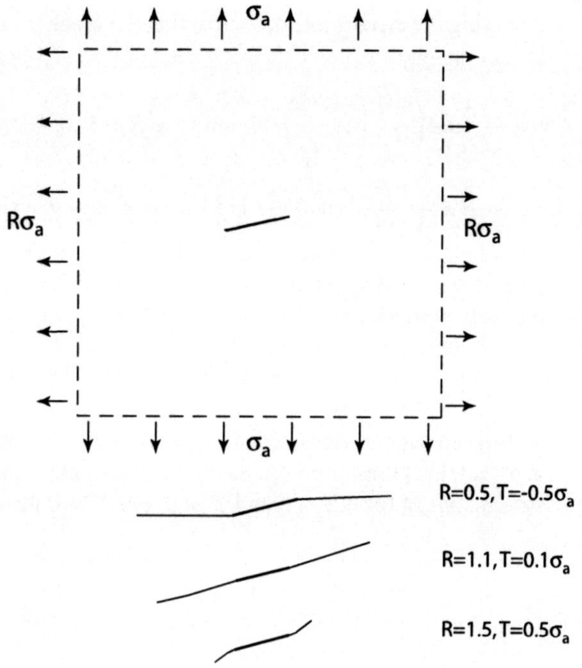

Fig. 4.13 Initially tilted crack under biaxial tension. Path shown for different ratios of σ_{22} to σ_{11}. Path computed using FRANC [8]

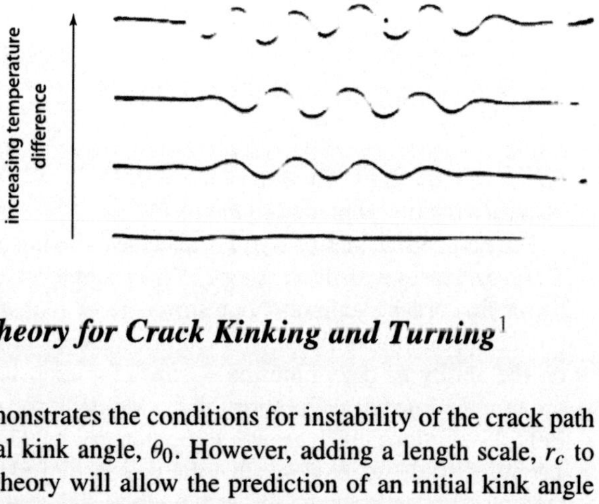

Fig. 4.14 As a heated glass slide is cooled from the left side the path transitions from straight to curved as the temperature difference increases. Further increases result in branching of the crack, adapted from [12]

4.4.4 Second Order Theory for Crack Kinking and Turning[1]

While the theory above demonstrates the conditions for instability of the crack path it does not predict the initial kink angle, θ_0. However, adding a length scale, r_c to the maximum hoop stress theory will allow the prediction of an initial kink angle even if $K_{II} = 0$. Adding the T stress the hoop stress at a distance r_c from the crack tip is

$$\sigma_{\theta\theta}(r_c, \theta) = \frac{1}{\sqrt{2\pi r_c}} \cos\frac{\theta}{2}\left(K_I \cos^2\frac{\theta}{2} - \frac{3}{2}K_{II}\sin\theta\right) + T\sin^2\theta. \tag{4.18}$$

[1]This section contributed by Jacob Hochhalter.

Applying the criterion (Eq. (4.10)) that the crack will kink at the angle θ^* that maximizes $\sigma_{\theta\theta}$ yields

$$\frac{1}{\sqrt{2\pi r_c}}\left(-\frac{3}{4}\cos\frac{\theta^*}{2}\right)(K_I\sin\theta^* + K_{II}(3\cos\theta^* - 1)) + 2T\sin\theta^*\cos\theta^* = 0$$
$$\frac{1}{\sqrt{2\pi r_c}}\left(-\frac{3}{8}(3\cos\theta^* - 1)\right)\left(K_I\cos\frac{\theta^*}{2} + K_{II}\sin\frac{\theta^*}{2}\right) + 2T\cos 2\theta^* < 0. \tag{4.19}$$

Notice the r_c value, which is not present in the original, 1st order, formulation of the $\max(\sigma_{\theta\theta})$ criteria,

$$K_I\sin\theta_c + K_{II}(3\cos\theta_c - 1) = 0.$$

First consider the case when $K_{II} = 0$. For $T < 0$ the maximum $\sigma_{\theta\theta}$ always occurs at $\theta^* = 0$ [13] thus as predicted by the crack path stability theory the crack will grow straight ahead for $K_{II} = 0$ and $T < 0$. For $T > 0$ the solution for the kink angle is

$$\theta^* = 0 \tag{4.20}$$

or

$$\theta^* = 2\cos^{-1}\left[\frac{1}{4\bar{T}} \pm \sqrt{0.5 + \frac{1}{(4\bar{T})^2}}\right] \tag{4.21}$$

where $\bar{T}$ is the normalized T stress,

$$\bar{T} = \frac{8}{3}\frac{T}{K_I/\sqrt{2\pi r_c}}. \tag{4.22}$$

For $\bar{T} < 1$ the only solution is $\theta^* = 0$. For $\bar{T} > 1$ there is a bifurcation and θ^* increases rapidly as plotted in Fig. 4.15.

For non-zero values of K_{II}/K_I the crack kinking angle increases with increasing $\bar{T}$ and decreases with decreasing $\bar{T}$. This change in kink angle with $\bar{T}$ is not captured by the first order maximum hoop stress theory (equivalent to $\bar{T} = 0$).

The application of this theory to finite element simulations of crack growth relies on the ability to determine the T stress for each increment of crack growth and on an experimental calibration of r_c. Methods for computing T are reviewed in Ref. [13]. Fatigue crack growth Experiments on 2024 aluminum alloy suggest $r_c \approx$ 1.5 mm [14]. Note, however that the presence of plastic deformation very close to the crack tip will change the actual stress fields rendering the above analysis an approximation for elastic-plastic materials.

4.5 Criteria for Fracture in Anisotropic Materials

Anisotropic materials may have fracture roughnesses that vary with crack orientation. Examples include layered materials such as mica or wood that have distinctly

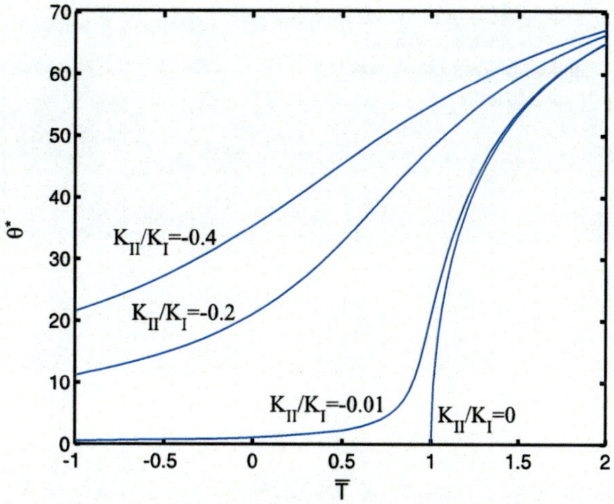

Fig. 4.15 Crack kinking angles predicted by second order maximum hoop stress theory. Kink angles are negative for positive K_{II}/K_I ratios

weaker bonding between the layers than along the in-plane directions and rolled metals that have slightly different toughnesses in different directions. When the toughness is anisotropic the criteria for fracture and for crack path selection must be modified.

The orientation dependent toughness of a material is defined as $G_C(\theta)$. This function could be smooth as in the case of rolled metals, or could be nearly constant with very low values only on specific planes as in the case of layered materials. For the maximum energy release rate criterion the fracture criterion is that the crack will grow at angle θ^* when [15]

$$G(\theta^*) \geq G_C(\theta^*) \quad \text{and} \quad \frac{\partial G(\theta^*)}{\partial \theta} = \frac{\partial G_C(\theta^*)}{\partial \theta}. \tag{4.23}$$

Thus, the crack may grow in a direction that is much different than the directions expected based on the loading. For example, layered materials loaded such that the principal stress is parallel to the layers may nonetheless fracture between the layers.

4.6 Crack Growth Under Fatigue Loading

Under repeated, or cyclic loading, materials can fail due to fatigue at stress levels well below their strength. Fatigue failure generally consists of three stages: (I) initiation of a crack (II) propagation of cracks and (III) final failure. The physical mechanisms for these stages will depend on the material and environmental conditions at hand. However in all cases, stage I will consist of the development of microstructural damage such a microcracks or slip bands. These will grow and eventually coalesce to form a dominant crack. Such a crack could be on the order of 1 mm long at the smallest. A great deal of the fatigue life of a component could be spent in stage I. In stage II the dominant crack grows stably under the application of repeated loads.

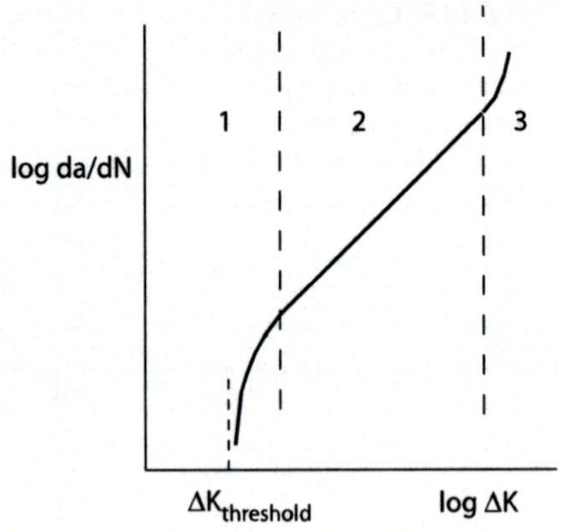

Fig. 4.16 Schematic of fatigue crack growth rate under constant amplitude cyclic loading

In stage III the crack has grown to a length where $K > K_{IC}$ and the component can fail unstably.

Fatigue life of structures is determined using total life or damage tolerant approaches. The total life approach predicts the fatigue life of a component as a total of the initiation and propagation time until failure. The damage tolerant design assumes that structures have imperfections and flaws from the beginning. Fatigue life is then calculated as the number of cycles needed to grow the crack to a size such that $K_I > K_{IC}$.

In stage II fatigue a crack will grow at values of K_I that are well below K_{IC}. Thus, since $r_p \sim (K_I/\sigma_y)^2$, the condition for small scale yielding is generally easily met. It is observed (particularly in metals) that the rate of crack growth per cycle of loading, da/dN is a function of the stress intensity factor range, $\Delta K = K_I^{max} - K_I^{min}$. Results of fatigue fracture tests are generally plotted on a log-log scale and will have a form similar to that sketched in Fig. 4.16. The data can be separated into three general regions. Below the threshold value at the left of region 1 there is no crack growth. Just above this threshold the crack grows very slowly. In region 2, at an intermediate value of ΔK the crack grows at a rate that can be expressed as a power law, [16]

$$\frac{da}{dN} = C\Delta K^n, \tag{4.24}$$

where C and n are experimentally determined parameters. Typically $2 < n < 4$. In region 3, the final stage of growth is marked by accelerating crack growth on the way to instability.

Actual data, Fig. 4.17, for example, show that the rate of fatigue crack growth depends not only on ΔK, but also on the "R-ratio", defined as $R \equiv P_{min}/P_{max}$. The dependence on R is due to "crack closure", or contact, of the crack faces when the load is removed [18, 19]. Crack closure can arise from many sources including plasticity in the wake of the growing crack, roughness of the fracture surface, oxidation of the new fracture surfaces and other effects, see [20, 21] and [22].

To understand the mechanics of crack closure, assume for now that the minimum of the cyclic load is zero. When of the load is removed from the component, the

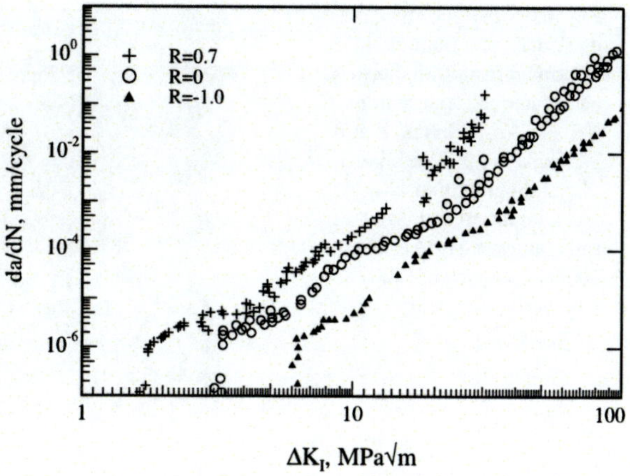

Fig. 4.17 Fatigue crack growth rate for 2024-T3 aluminum alloy, 2.3 mm thick, at three R ratios, $R = -1.0$, $R = 0$ and $R = 0.7$ in moist air. Note that da/dN spans 7 orders of magnitude. Adapted from [17]

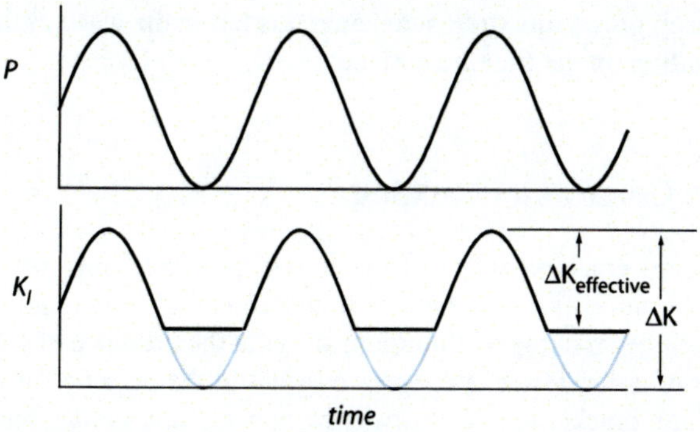

Fig. 4.18 Crack closure effect. In this example load ratio is $R = 0$. The nominal value of the stress intensity factor amplitude, ΔK is calculated based on the applied load P. However, due to contact behind the crack the actual stress intensity factor does not fall to 0 when $P = 0$, thus the actual, or "effective" stress intensity factor $\Delta K_{\text{effective}}$ is less than the nominal value. As R increases, the minimum value of K_I stays the same and thus the reduction in $\Delta K_{\text{effective}}$ due to closure becomes less. Thus, when da/dN is plotted vs. ΔK, the rate of growth is higher for higher R. When $R > 0.7$ the effect of closure largely disappears

crack faces close and can go into compressive contact with each other, partially holding the crack open. This maintains a non-zero K_I at the crack tip, as shown in Fig. 4.18. Thus in a cycle of loading the amplitude of the stress intensity factor variation, ΔK is reduced relative to the case in which no closure occurs,reducing the crack growth rate relative to the case in which no closure occurs. As R increases to above approximately 0.7 the cracks do not contact even at the minimum of the load and closure does not occur.

Note as well that environmental conditions can play an important part in fatigue crack growth. For example fatigue tests on aluminum alloys show that the crack growth rate is faster in humid air than in inert environments.

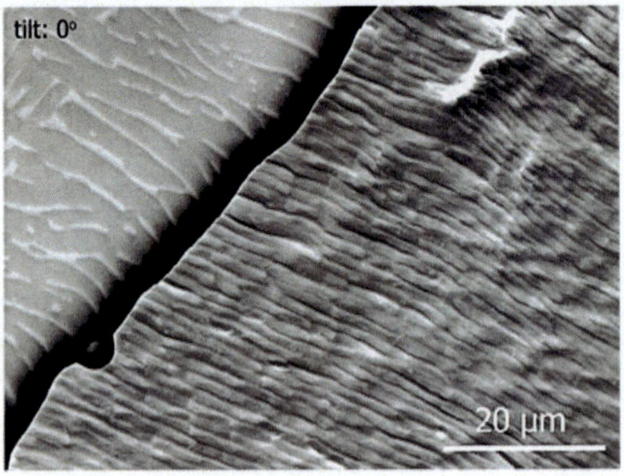

Fig. 4.19 Fatigue striations on surface of a titanium alloy. Reprinted from Metallurgical and Materials Transactions A Vol. 41, A.L. Pilchak, R.E.A. Williams and J.C. Williams, "Crystallography of fatigue crack initiation and growth in fully lamellar Ti-6Al-4V," 106-124, copyright (2009), with permission from Springer [23]

The fracture surfaces of fatigue cracks can often be distinguished by striations as shown in Fig. 4.19. During loading, the material ahead of the crack is stretched. When the crack grows this stretched material is left in the wake of the crack, leaving behind a striation on the fracture surface.

4.7 Stress Corrosion Cracking

Stress corrosion cracking refers to the time-dependent, slow growth of cracks in corrosive environments under the combined effects of stress and chemical attack. Examples include cracking of aluminum alloys in the presence of salt water, steels in the presence of chlorides or hydrogen and glass in the presence of water. In metals, stress corrosion cracks will typically grow between the grains, but may also grow across grains. Stress corrosion cracking is time dependent with the rate of growth depending on both the stress intensity factor and the corrosive environments. For example the rate of crack growth in a high-strength steel alloy (4340) in the presence of water and hydrogen is shown in Fig. 4.20(a) [24].

The rate of crack growth in high strength AISI 4335V steel in the presence of salt water, hydrogen and hydrogen sulfide is shown in Fig. 4.20(b) [24]. Typically below a threshold level of K_I, no cracking occurs, and as the stress intensity factor reaches K_{IC} the crack grows rapidly. In the middle region the rate of growth depends on the availability of hydrogen. However, such plateau regions are not found in all materials. Stress corrosion cracking is a significant issue in structures of many types. Much more information on this topic can be found in [24] and [25].

4.8 Exercises

1. In [9] it is shown that for a straight crack with a kink of infinitesimal length the stress intensity factors at the tip of the kink can be calculated from

$$K_I = C_{11}k_I + C_{12}k_{II},$$

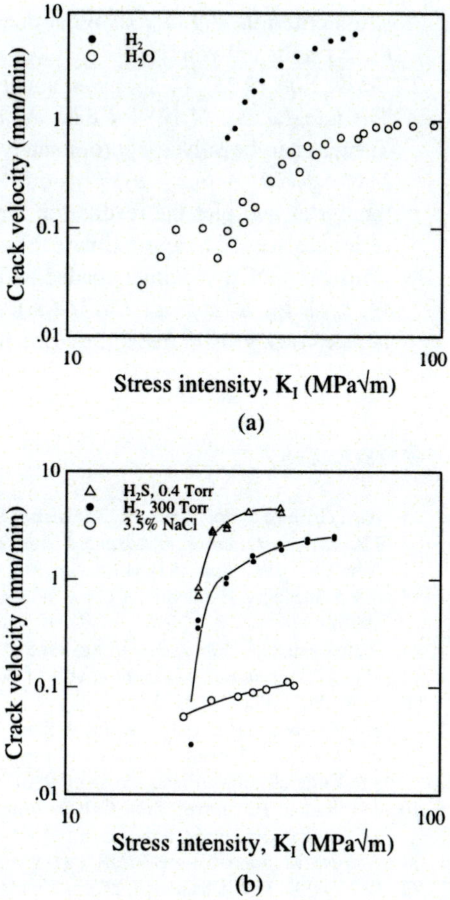

Fig. 4.20 Stress corrosion cracking rates for (**a**) 4340 steel and (**b**) AISA 4335V steel. Adapted from [24]

$$K_{II} = C_{21}k_I + C_{22}k_{II}, \tag{4.25}$$

where

$$\begin{aligned}
C_{11} &= \frac{1}{4}\left(3\cos\alpha/2 + \cos 3\alpha/2\right),\\
C_{12} &= -\frac{3}{4}\left(\sin\alpha/2 + \sin 3\alpha/2\right),\\
C_{21} &= \frac{1}{4}\left(\sin\alpha/2 + \sin 3\alpha/2\right),\\
C_{22} &= \frac{1}{4}\left(\cos\alpha/2 + 3\cos 3\alpha/2\right).
\end{aligned} \tag{4.26}$$

Using the criterion that the crack will grow at an angle that holds $K_{II} = 0$, calculate and plot the kink angle θ^* vs. $0 \le \beta \le 90°$ for a crack loaded as shown in Fig. 4.9. Compare the predicted kink angle to the results of the maximum hoop stress criterion, Fig. 4.10a. What does the disagreement or agreement between these results tell you?

2. Derive Eqs. (4.25), (4.26) given above.
3. The integral Eq. (4.16) for the path of a perturbed crack in the presence of a T stress can be solved approximately by expressing the path, $\lambda(x)$ by a series, $\lambda(x) = b_0 x^{1/2} + b_1 x + b_2 x^{3/2} + b_4 x^2 + b_5 x^{5/2} + \cdots$. Evaluate the constants in this series and plot the results for various values of η. How many terms in the series are needed to approximate the results in Fig. 4.12?
4. Consider a DCB specimen loaded with a compliant testing machine. Plot G/G_C vs. a/a_0 for a range of $C_M/C(a_0)$. How soft must the loading system be to induce crack growth instability for a material with a flat R-curve?

References

1. A.A. Griffith, Philos. Trans. R. Soc. Lond. Ser. A, Math. Phys. Sci. **221**, 163 (1921)
2. J.R. Rice, in *Fracture an Advanced Treatise*, vol. 2, ed. by H. Liebowitz (Academic Press, New York, 1968), pp. 191–311, Chap. 3
3. ASM Specialty Handbook, *Carbon and Alloy Steels* (ASM International, Materials Park, 1996)
4. F. Erdogan, G.C. Sih, J. Basic Eng. **85**, 516 (1963)
5. G.C. Sih, *Mechanics of Fracture*, vol. 1 (1972), Chap. 1
6. C.W. Wu, J. Elast. **8**, 2235 (1978)
7. R.V. Goldstein, R.L. Salganik, Int. J. Fract. **10**, 507 (1974)
8. http://www.cfg.cornell.edu/software/software.htm. Last accessed January 3, 2008
9. J.R. Cotterell, Int. J. Fract. **16**, 155 (1980)
10. J.C. Radon, P.S. Lever, L.E. Culver, *Fracture 1977*, vol. 3 (University of Waterloo Press, 1977), chap. Fracture toughness of PMMA under biaxial stress, pp. 1113–1118
11. A. Yuse, M. Sano, Nature **362**, 329 (1993)
12. B. Yang, K. Ravi-Chandar, J. Mech. Phys. Solids **49**, 91 (2001)
13. C.S. Chen, Crack growth simulation and residual strength prediction in thin shell structures. Ph.D. thesis, Cornell University (1999)
14. R.G. Pettit, J.C. Newman, M.S. Domack, *19th ICAF Symposium* (1997)
15. Palaniswamy, W. Knauss, Mech. Today **4**, 87 (1978)
16. P. Paris, F. Erdogan, J. Basic Eng. **85**, 528 (1963)
17. R.P. Gangloff, R.S. Piascik, D.L. Dicus, J.C. Newman Jr., J. Aircr. **31**(3), 720 (1994)
18. W. Elber, Eng. Fract. Mech. **2**, 37 (1970)
19. W. Elber, in *Damage Tolerance in Aircraft Structures*. Elastic-Plastic Fracture, vol. 486 (American Society for Testing and Materials, Philadelphia, 1971), pp. 230–242
20. S. Suresh, *Fatigue of Materials*, 2nd edn. (Cambridge University Press, Cambridge, 1998)
21. J.C. Newman, in *Elastic-Plastic Fracture*, vol. 1343, ed. by R.C. McClung, J.C. Newman (American Society for Testing and Materials, 2000), pp. 128–144
22. J.A. Newman, R.S. Piascik, Int. J. Fatigue **26**, 923 (2004)
23. A. Pilchak, R. Williams, J. Williams, Metall. Mater. Trans. A, Phys. Metall. Mater. Sci. **41**, 106 (2010)
24. G.E. Kerns, M.T. Wang, R.W. Staehle, in *Stress Corrosion Cracking and Hydrogen Embrittlement of Iron Base Alloys*, ed. by J. Hochmann, R.W. Staehle (National Association of Corrosion Engineers, 1973), pp. 700–734
25. L.L. Shreir, R.A. Jarman, G.T. Burstein, *Corrosion*, vol. 1 (Butterworth/Heinemann, Stoneham, 2000), Chap. 8, Effect of Mechanical Factors on Corrosion

Chapter 5
Determining K and G

Abstract So far the crack tip stress fields and energy flows for elastic materials have been described and we have shown that near the crack tip the stress, strain and displacement fields are completely determined by $\{K_I, K_{II}, K_{III}\}$. Furthermore the use of stress intensity factors and energy release rate as criteria for fracture have been introduced. However for only a small number of cases have solutions for the stress intensity factors been given. Thus in this chapter analytical, look-up, computational and experimental methods for the determination of the stress intensity factors and energy release rate will be described.

5.1 Analytical Methods

5.1.1 Elasticity Theory

The equations of 2D elasticity can be formulated in terms of complex variable theory, thus allowing the powerful methods of analytic function theory and conformal mapping to be applied. With such methods the solutions to a great number of fracture problems have been found. A brief outline of the method is given in Sect. 2.6 and results for two cases, that of a crack in an infinite plate under tensile loading and a crack subject to crack face tractions. Further examples a description of the method can be found in [1] and [2]. Analytical solutions are useful not only to calculate stress intensity factors for physical problems that can be approximated by these idealizations, but as building blocks for more complex solutions and as examples against which to test computational methods for calculating stress intensity factors.

5.1.1.1 Finite Crack in an Infinite Body

The case of an anti-plane shear crack under uniform, remote stress is given in Fig. 2.4. The stress intensity factors for a finite crack under uniform remote tensile or shear stress are given in Sect. 2.6.4.

Summarizing the results given in Sect. 2.6.6 and adding results for anti-plane shear, if the crack surfaces have non-uniform traction loading $\mathbf{t} = p_1(x_1)\mathbf{e}_1 +$

A.T. Zehnder, *Fracture Mechanics*,
Lecture Notes in Applied and Computational Mechanics 62,
DOI 10.1007/978-94-007-2595-9_5,

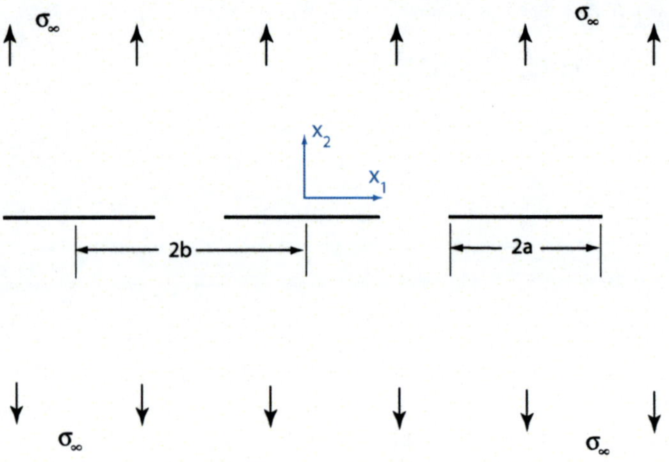

Fig. 5.1 Array of cracks of length $2a$ and spacing $2b$ under uniform, remote tension load

$p_2(x_1)\mathbf{e}_2 + p_3(x_1)\mathbf{e}_3$ on the top surface and $\mathbf{t} = -p_i(x_1)\mathbf{e}_i$ on the bottom surface, the stress intensity factors are

$$\begin{Bmatrix} K_I \\ K_{II} \\ K_{III} \end{Bmatrix} = \frac{1}{\sqrt{\pi a}} \int_{-a}^{a} \left[\frac{a+t}{a-t}\right]^{1/2} \begin{Bmatrix} p_2(t) \\ p_1(t) \\ p_3(t) \end{Bmatrix} dt. \tag{5.1}$$

5.1.1.2 Semi-infinite Crack in an Infinite Body

The stress intensity factors for a semi-infinite crack, Fig. 2.3, loaded by tractions are

$$\begin{Bmatrix} K_I \\ K_{II} \\ K_{III} \end{Bmatrix} = \frac{\sqrt{2}}{\sqrt{\pi}} \int_{-\infty}^{0} \frac{1}{\sqrt{-t}} \begin{Bmatrix} p_2(t) \\ p_1(t) \\ p_3(t) \end{Bmatrix} dt. \tag{5.2}$$

The crack opening displacement for the case of Mode-I loading is

$$u_2^+(x_1) - u_2^-(x_1) = \frac{4}{\pi E'} \int_{-\infty}^{0} p_2(t) \ln \left| \frac{\sqrt{-x_1} + \sqrt{-t}}{\sqrt{-x_1} - \sqrt{-t}} \right| dt, \quad x_1 < 0. \tag{5.3}$$

5.1.1.3 Array of Cracks Under Remote Loading

For a sheet perforated with an array of cracks, each of length $2a$, center-to-center spacing $2b$, and loaded with uniform, far-field tension, as shown in Fig. 5.1,

$$K_I = \sigma_\infty \sqrt{\pi a} \left[\frac{2b}{\pi a} \tan \frac{\pi a}{2b}\right]^{1/2}. \tag{5.4}$$

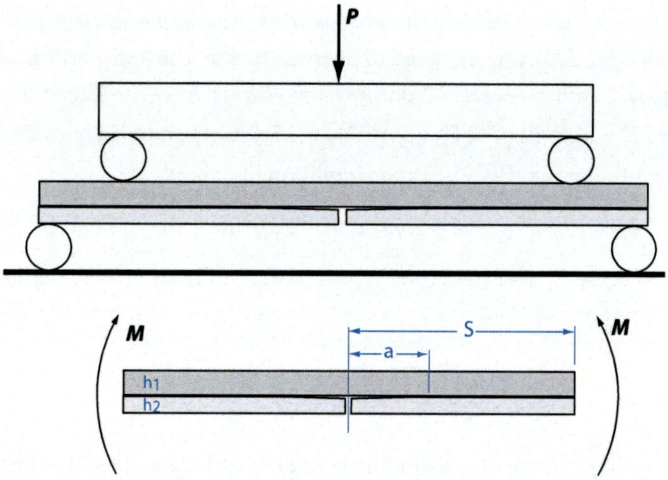

Fig. 5.2 Two beams bonded together and loaded in 4-point bending. The section between the inner loading pins is in pure bending with moment M [3]

5.1.2 Energy and Compliance Methods

The energy and compliance methods, introduced in Chap. 3 are useful in the case where a test specimen or structure can be modeled using beam or plate theory. In such cases if the energy or stiffness of the structure can be determined as a function of crack length, or area, then the energy release rate and (if the loading is Mode-I) stress intensity factor can be computed. A few examples are given here.

5.1.2.1 4-Point Bending Debond Specimen: Energy Method

The 4-point bending sample consisting of two bonded beams of thicknesses h_1 and h_2, shown in Fig. 5.2 is useful for measuring delamination and debonding fracture toughness. The energy release rate for this sample can easily be computed, however, since the cracks are subject to mixed-mode loading, further analysis is needed to determine the individual stress intensity factors [3].

The portion of the test sample between the inner loading pins is loaded in pure bending with a moment M. Treating this as a fixed force problem the energy release rate can be calculated from Eq. (3.17). Assuming that the material is linearly elastic, $\Omega = \frac{1}{2}Pq$ and thus Eq. (3.17) can be written as $G = \frac{\partial \Omega}{\partial s}$. If B is the depth of the beam sample, then $s = 2B$ (the factor of two accounts for the two cracks) and

$$G = \frac{1}{2B}\frac{\partial \Omega}{\partial a} \tag{5.5}$$

where a is the crack length. In pure bending the strain energy per unit length is $\frac{M^2}{2EI}$, where I is the moment of inertia. In the cracked portion of the beam $I = Bh_2^3/12$. In the uncracked portion $I = B(h_1 + h_2)^3/12$. Since the moment is constant, $\Omega = \frac{1}{2}\frac{M^2}{E}[\frac{2a}{Bh_1^2/12} + \frac{2(S-a)}{B(h_1+h_2)^3/12}]$. Taking the derivative of Ω with respect to a, substituting into Eq. (5.5) and simplifying,

$$G = \frac{6M^2}{EB^2}\left[\frac{1}{h_1^3} - \frac{1}{(h_1 + h_2)^3}\right]. \tag{5.6}$$

5.2 Stress Intensity Handbooks and Software

Using a variety of methods, including boundary collocation, energy approaches and conformal mapping solutions for stress intensity factors for a great many practical problems have been calculated by numerous researchers over the years. In an effort spanning decades these results have been tabulated in easy to use, well organized handbooks [4, 5]. Generally these handbooks provide equations for stress intensity factors as a function of the geometry and dimensions of the crack and of the object containing the crack. The results are given a graphs, equations and tables of coefficients. A sampling of stress intensity factor solutions for common fracture test specimens is given in Table 5.1.

Taking the use of tabulated solutions further, software packages such as NASCRAC [6] and NASGRO [7] integrate stress intensity factor solutions, material property databases and a graphical user interface to provide tools for the estimation of allowable loads, fatigue life and other calculations of interest in practical applications.

5.3 Boundary Collocation

Boundary collocation is a method for the computation of stress intensity factors based on the eigenfunction expansion of the crack tip stress fields. The general idea is that given traction boundary conditions for a certain problem, express the tractions at a finite number of locations, called collocation points, in terms of the eigenfunction expansion with unknown coefficients. For each collocation point you will have one equation for the unknown coefficients. as long as you use at least as many collocation points as unknowns in the expansion, you can then solve the resulting system of equations for the eigenfunction coefficients. The value of the coefficient corresponding to the $r^{-1/2}$ stress singularity is the stress intensity factor.

Although this method may be applied to Mode-I, II or III problems, here let us consider (to make it simpler) Mode-III problems. Recall that in the complex variables formulation for anti-plane shear

Table 5.1 Stress intensity solutions for several fracture test specimen geometries. $E' = E$ (plane stress), $E' = E/(1-\nu^2)$ (plane strain). Adapted from [4] and [8]

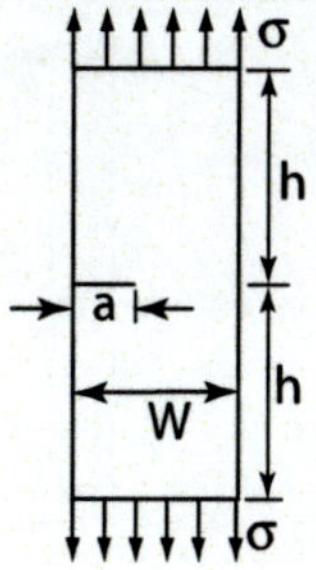

Single Edge Notch Tension (SENT)

$h/W > 1$

$K_I = \sigma\sqrt{\pi a}F(a/W)$

$F(a/W) = 0.265(1-a/W)^4 + \frac{0.857+0.265a/W}{(1-a/W)^{3/2}}$

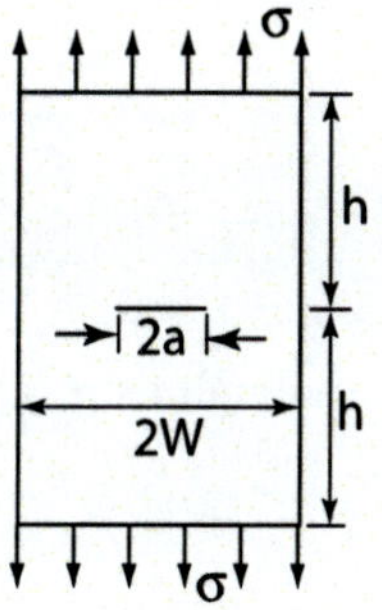

Center Cracked Tension (CCT)

$h/W > 3$

$K_I = \sigma\sqrt{\pi a}F(a/W)$

$F(a/W) = \sqrt{\sec\frac{\pi a}{2W}}[1 - 0.025(a/W)^2$
$+ 0.06(a/W)^4]$

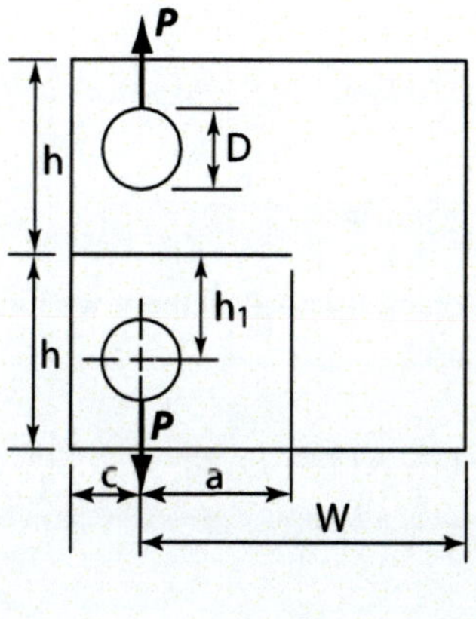

ASTM Standard Compact Tension (CT)

$h = 0.6W$, $h_1 = 0.275W$, $D = 0.25W$, $c = 0.25W$, thickness, $B = W/2$

$K_I = \frac{P}{WB}\sqrt{a}F(a/W)$

$F(a/W) = 29.6 - 185.5(a/W) + 655.7(a/W)^2$
$- 1017.(a/W)^3 + 638.9(a/W)^4$

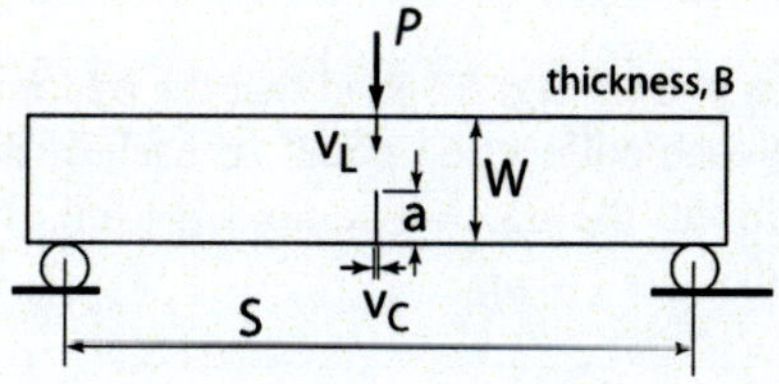

Three Point Bend (Single Edge Notch Bend, SENB) $S/W = 4$, $K_I = \frac{PS}{BW^{3/2}}F(a/W)$

$F(a/W) =$

$$\frac{3(a/W)^{1/2}\left[1.99-(a/W)(1-a/W)(2.15-3.93a/W+2.7(a/W)^2)\right]}{2(1+2a/W)(1-a/W)^{3/2}}$$

Load-line disp, $v_L = \frac{P}{E'B}(\frac{S}{W-a})^2[1.193 - 1.98a/W$
$+ 4.478(a/W)^2 - 4.443(a/W)^3 + 1.739(a/W)^4]$

CMOD, $v_C = \frac{4\sigma a}{E'}[0.76 - 2.28a/W + 3.87(a/W)^2$
$- 2.04(a/W)^3 + 0.66/(1-a/W)^2]$, $\sigma = \frac{6PS}{4BW^2}$

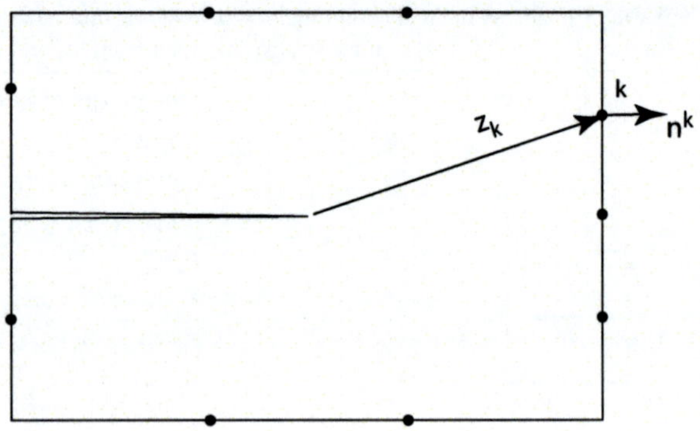

Fig. 5.3 Boundary collocation points, $k = 1, m$ points ($m = 9$ here)

$$\tau = \sigma_{32} + i\sigma_{31},$$
$$n = n_1 + in_2,$$
$$t = t_3 = \mathrm{Im}(\tau n),$$

where t_3 is the x_3 component of traction. It can be shown that the complex stress, τ can be expanded in a Laurent series,

$$\tau = \sum_{n=-\infty}^{+\infty} A_n z^{\lambda_n}.$$

To satisfy traction free boundary conditions, $\mathrm{Re}(\tau) = 0$ (σ_{32}) on $\theta = \pm\pi$, or

$$\mathrm{Re}(\tau(r,\pi)) = \sum A_n r^{\lambda_n} \cos\lambda_n\pi = 0 \quad \Rightarrow \quad \lambda_n = \frac{-1}{2} + n.$$

For a sharp crack we argued that to prevent infinite crack tip displacement we must eliminate any λ_n less than $-\frac{1}{2}$, so that for a *sharp crack*,

$$\tau = \sum_{n=0}^{+\infty} A_n z^{\lambda_n}, \qquad \lambda_n = -\frac{1}{2} + n. \tag{5.7}$$

In this case, $A_0 = \frac{K_{III}}{\sqrt{2\pi}}$.

Assume that we have $k = 1, m$ collocation points, Fig. 5.3, and that the traction, t_3^k, position, z_k and normal n^k are known at each collocation point. At each of the m collocation points equate the given traction to the traction computed using the asymptotic expansion containing $p+1$ points, $p + 1 \leq m$,

$$\text{for } k = 1, m$$

$$\mathrm{Im}\left(\sum_{n=0}^{p} A_n z_k^{(-\frac{1}{2}+n)} n^k\right) = t_3^k. \tag{5.8}$$

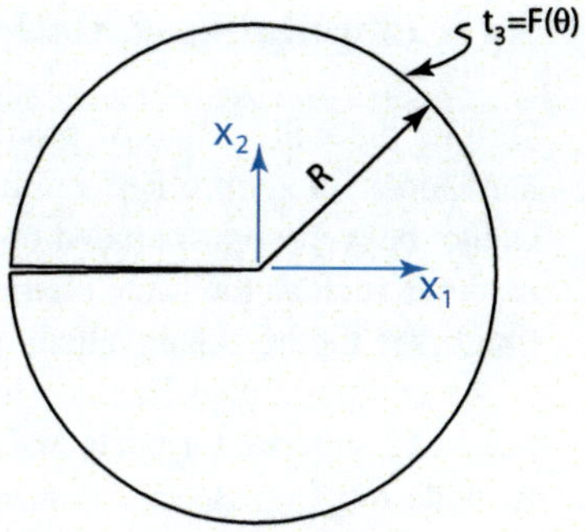

Fig. 5.4 Crack in circular cylinder of radius R. Cylinder is loaded with tractions $t = F(\theta)$

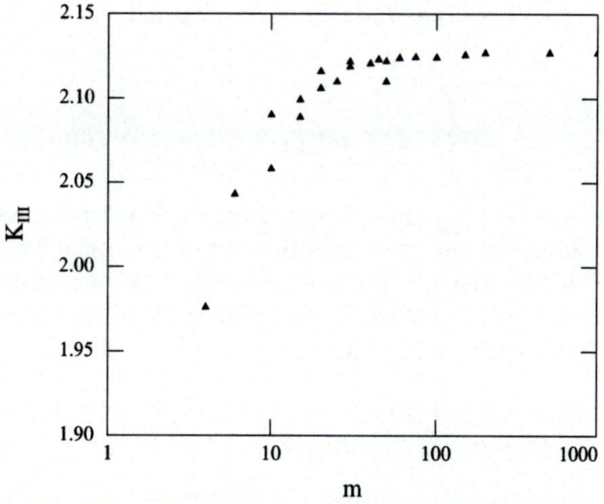

Fig. 5.5 Results from boundary collocation calculation for $F(\theta) = \sin\theta$

This sets up a system of m equations for $p + 1$ unknowns, $A_0, \ldots, A_p$. The value of the leading term, A_0 corresponds to the stress intensity factor, $K_{III} = \sqrt{2\pi} A_0$.

For best accuracy more collocation points than unknowns should be used; as a rule of thumb take $m \approx 2(p + 1)$ or greater. In this case the system of Eqs. (5.8) must be solved in the least squares sense. You can experiment with the number of collocation points and number of terms in the eigenfunction expansion needed to provide good accuracy. Generally the more terms and the more collocation points the better your accuracy will be. As an example consider the case of a crack embedded in a circular region of radius $R = 1$ as shown in Fig. 5.4. The cylinder is loaded with tractions $t_3 = F(\theta)$. Collocation points were placed at equal angles around the cylinder. Varying the number of collocation points and the number of terms in the eigenfunction expansion, the system of Eqs. (5.8) was solved and the stress intensity factor determined. Results for the case $F(\theta) = \sin\theta$ are shown in Fig. 5.5 for various values of m, and varying the number of terms in the eigenfunction expansion. The results show that for $m \approx 20$ or greater that the value of K_{III} has essentially converged.

5.4 Computational Methods: A Primer

Despite the wide range of problems which have been solved and tabulated in the handbooks, many problems in fracture mechanics involve complex geometries that cannot be well approximated by the handbook solutions. The use of computational methods such as the finite element method, boundary element method and dislocation based methods is invaluable for studying fracture in real-world problems.

As computational fracture is itself a vast field, here we will only study some basic but important aspects and we will focus on the finite element method due to its wide ranging use in engineering design and since it is a very flexible method, extendable to nonlinear and dynamic problems. The emphasis here is on methods that can be used with standard finite element packages and that require relatively simple post-processing. For further details see [9, 10].

5.4.1 *Stress and Displacement Correlation*

Since the asymptotic crack tip stress and displacement fields are completely determined by the stress intensity factors, one should be able to extract stress intensity factors from the results of a finite element simulation of a cracked body by matching the asymptotic fields to the computed stress and displacement fields [11]. This is the idea behind the stress and displacement correlation methods.

5.4.1.1 Stress Correlation

Recall that the stress intensity factors are defined as

$$\begin{pmatrix} K_I \\ K_{II} \\ K_{III} \end{pmatrix} = \lim_{r \to 0} \sqrt{2\pi r} \begin{pmatrix} \sigma_{22}(r,0) \\ \sigma_{12}(r,0) \\ \sigma_{32}(r,0) \end{pmatrix}. \tag{5.9}$$

If the stress along $\theta = 0$ can be calculated, then the stress intensity factor can be determined by extrapolation back to $r = 0$ [11]. For example, in Mode-I, use the FEM method to compute the stresses at points ahead of the crack tip, then plot $K_I = \lim_{r\to 0} \sigma_{22}(r,0)\sqrt{2\pi r}$ vs. r. Extrapolate this curve back to $r = 0$ to determine K_I. Advantages of this method are that it is quite simple, it can be used with any finite element program, no special postprocessing is needed, only one analysis is needed, different modes of stress intensity factors are easily computed and stress intensity factors can be computed along a 3D crack front by taking stresses on lines normal to the crack front at different positions. The accuracy of the method will depend on mesh refinement and the ability of the mesh to capture the crack tip stress singularity. The method is easily generalized to use the stresses on any angle θ, which may be desirable since, depending on the mesh used, the integration points may not lie on $\theta = 0$.

5.4.1.2 Displacement Correlation

A similar approach can be taken based on the displacements of the crack face. Recall that the displacements along the crack line ($\theta = \pi$) are given by

$$\begin{pmatrix} u_1(r,\pi) \\ u_2(r,\pi) \\ u_3(r,\pi) \end{pmatrix} = \sqrt{\frac{r}{2\pi}} \begin{pmatrix} -\frac{4K_{II}}{E'} \\ \frac{4K_I}{E'} \\ \frac{K_{III}}{\mu} \end{pmatrix}. \tag{5.10}$$

Using the finite element method calculate the displacements along the crack line then (for Mode-I, for example) plot $K_I^* = E'\sqrt{2\pi}u_2/4\sqrt{r}$ vs. r and extrapolate the plot back to $r = 0$. This has the advantage over stress correlation that the displacements are primary solution variables in the finite element method. Similarly to stress correlation different modes of stress intensity factor can be calculated and stress intensity factors along a 3D crack front can be calculated. Accuracy issues are similar to stress correlation.

5.4.2 Global Energy and Compliance

Recall that $G = J$ and that for cracks that grow straight ahead, $J = \frac{K_I^2}{E'} + \frac{K_{II}^2}{E'} + \frac{K_{III}^2}{2\mu}$. Thus if the problem involves only a single mode of loading, then the stress intensity factor can be extracted by finding the energy release rate.

Recall that for a 2D crack in a linear elastic material $G = -\frac{\partial \Omega}{\partial a}$ for fixed displacement loading and $G = \frac{\partial \Omega}{\partial a}$ for fixed force loading. Thus one method to determine G is to perform two analyses, one for a crack of length a and another for a crack of length $a + \Delta a$, find the total strain energy for each problem and calculate G from

$$G \approx \pm \frac{\Omega(a + \Delta a) - \Omega(a)}{\Delta a}, \tag{5.11}$$

using + or − appropriate to the loading type. Note that if a unit thickness is used in the FEM calculation then Ω will be energy/thickness and Δa in the above is the increment in crack length. If a unit thickness is not used in the calculation, or if a 3D calculation were being performed, then Δa in Eq. (5.11) must be replaced with the *area* of the crack increment.

Similarly the stress intensity factor could be calculated using the relation between compliance and G, $G = \frac{1}{2}P^2\frac{\partial C}{\partial a}$, where P is the generalized load. The change in compliance $\frac{\partial C}{\partial a} \approx \frac{C(a+\Delta a) - C(a)}{\Delta a}$ could be found by using two finite element calculations as with the energy method.

Advantages of these global approaches are that they are not as sensitive as stress and displacement correlation to the crack tip meshing since they deal with global quantities and no special finite element code or postprocessing is needed. Disadvantages are that two calculations are needed, individual stress intensity factor modes cannot be determined and stress intensity factor variations along a crack front cannot be determined.

5.4.3 Crack Closure Integrals

The crack closure integral, Eq. (3.38) for 2D cracks that grow straight ahead, can be rewritten in terms of energy release rates for each mode, each uniquely related to a stress intensity factor, i.e.,

$$\begin{aligned}
G &= G_1 + G_2 + G_3 \\
&= \frac{K_I^2}{E'} + \frac{K_{II}^2}{E'} + \frac{K_{III}^2}{2\mu}, \\
G_1 &= \lim_{\Delta a \to 0} \frac{1}{\Delta a} \int_0^{\Delta a} \sigma_{22}(x_1, 0) u_2(\Delta a - x_1, \pi)\, dx_1, \\
G_2 &= \lim_{\Delta a \to 0} \frac{1}{\Delta a} \int_0^{\Delta a} \sigma_{12}(x_1, 0) u_1(\Delta a - x_1, \pi)\, dx_1, \\
G_3 &= \lim_{\Delta a \to 0} \frac{1}{\Delta a} \int_0^{\Delta a} \sigma_{32}(x_1, 0) u_3(\Delta a - x_1, \pi)\, dx_1.
\end{aligned} \tag{5.12}$$

The crack closure integral can be approximated in a finite element computation and used to extract individual stress intensity factors.

5.4.3.1 Nodal Release

In the nodal release (NR) method [12] two analyses are performed, one for crack of length a and another for length $a + \Delta a$. Consider the mesh shown in Fig. 5.6 with node j at the crack tip in the first configuration with crack length a. Let F_i be the components of the nodal reaction force at node j in the first configuration (before crack “released”) and let $\Delta u_i = u_i^{(+)} - u_i^{(-)}$ be the components of the crack face opening or sliding displacements at node j in configuration two (after the crack tip node is released). The energy release rate components can be calculated from

$$\begin{aligned}
G_1 &= \frac{1}{\Delta a} F_2 \frac{\Delta u_2}{2}, \\
G_2 &= \frac{1}{\Delta a} F_1 \frac{\Delta u_1}{2}, \\
G_3 &= \frac{1}{\Delta a} F_3 \frac{\Delta u_3}{2},
\end{aligned} \tag{5.13}$$

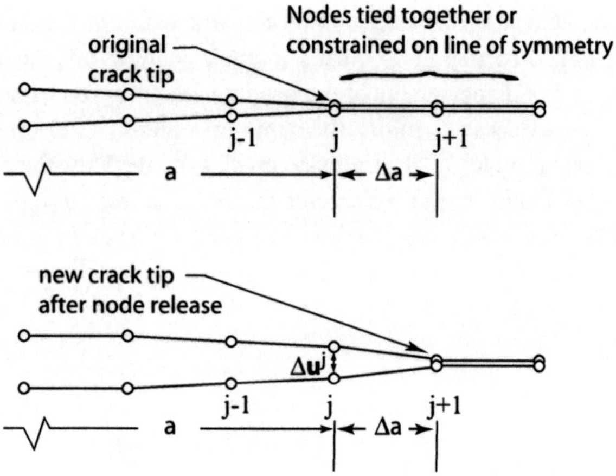

Fig. 5.6 Schematic of nodal release procedure for calculation of energy release rates

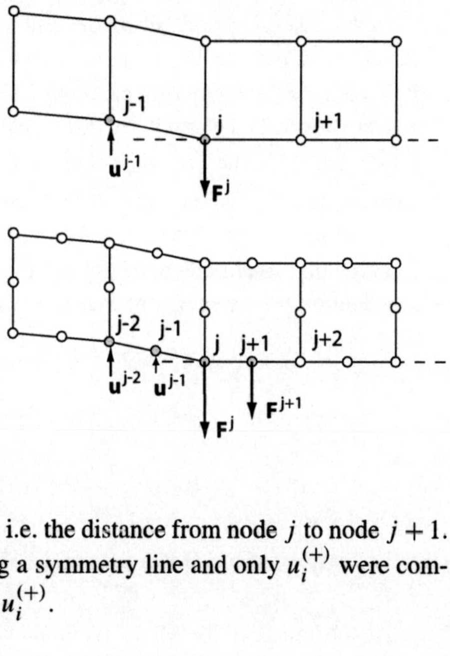

Fig. 5.7 Schematic of MCCI procedure for computing energy release rates

where Δa is the length of crack growth, i.e. the distance from node j to node $j+1$. Note that if the nodes were placed along a symmetry line and only $u_i^{(+)}$ were computed, replace $\Delta u_i/2$ in the above with $u_i^{(+)}$.

5.4.3.2 Modified Crack Closure Integral

A similar approach is the modified crack closure integral (MCCI) [13–15]. Consider the two crack tip meshes shown in Fig. 5.7, one using linear, 4 node elements and the other with quadratic, 8 node elements. In both meshes the element lengths along the x_1 direction are uniform and equal to Δa. The crack tip mesh must be sufficiently refined so that for a crack growing over the distance of one element the crack growth is self-similar, i.e. $K(a+\Delta a) \approx K(a)$.

Consider first the 4 node element with crack tip node j. If the crack were to advance by one node, then the crack opening displacement for the new crack would be the displacement at the original crack tip, i.e. at node j. Assuming that the crack growth is self-similar, the opening at node j after crack growth is equal to the opening at node $j-1$ in prior to crack growth. Thus the energy release rate components for linear (4 node) elements can be calculated as

$$
\begin{aligned}
G_1 &= \frac{1}{2\Delta a} F_2^j \Delta u_2^{j-1}, \\
G_2 &= \frac{1}{2\Delta a} F_1^j \Delta u_1^{j-1}, \\
G_3 &= \frac{1}{2\Delta a} F_3^j \Delta u_3^{j-1},
\end{aligned}
\tag{5.14}
$$

where $\Delta u_i^{j-1} = u_i^{(+)j-1} - u_i^{(-)j-1}$.

In the case of 8 node elements with crack tip node j, when the crack grows by one element, from node j to $j+2$, then the crack opening corresponding to the force F_i^j which is released during crack growth, is the displacement two nodes behind the new crack tip. If the crack growth is self-similar, this is equal to the displacement at two nodes behind the original crack tip, i.e. equal to Δu_i^{j-2}. The crack opening displacement corresponding to the release of the force F_i^{j+1} is the displacement one node behind the new crack tip. Again, if crack growth is self-similar this is equal to the displacement one node behind the original crack tip, i.e. Δu_i^{j-1}. Putting the contributions to the energy release rate together, the energy release rate for quadratic (8 node) elements is

$$
\begin{aligned}
G_1 &= \frac{1}{2\Delta a}(F_2^j \Delta u_2^{j-2} + F_2^{j+1} \Delta u_2^{j-1}), \\
G_2 &= \frac{1}{2\Delta a}(F_1^j \Delta u_1^{j-2} + F_1^{j+1} \Delta u_1^{j-1}), \\
G_3 &= \frac{1}{2\Delta a}(F_3^j \Delta u_3^{j-2} + F_3^{j+1} \Delta u_3^{j-1}).
\end{aligned}
\tag{5.15}
$$

An advantage of the MCCI method over NR is that only one analysis is required. In addition, as with NR, only very simple post-processing of the data, namely extraction of the nodal forces and displacements are needed. Note that in both the NR and MCCI methods if the crack is not a line of symmetry, then the elements above and below the line ahead of the crack must have separate nodes. These nodes can then be constrained to have equal displacement so that nodal reaction forces can be extracted from the analysis. Note as well that if the crack did not grow straight ahead then the equations above for G would not be valid. Generalizations of MCCI have been developed for use with quarter point and other elements [16, 17] where the simple relations above will not work.

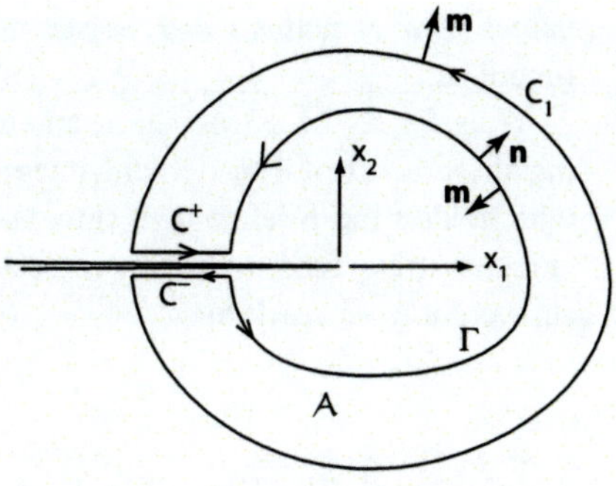

Fig. 5.8 Contours for derivation of domain integral calculation of J

5.4.4 Domain Integral

Perhaps the most accurate and elegant method for computing the energy release rate is to calculated the J integral by converting the line integral into a domain integral which can easily be calculated using the known finite element shape functions [18–20]. Consider the closed contour $\mathscr{C}$ with outward unit normal vector $\mathbf{m}$ as shown in Fig. 5.8. The contour $\mathscr{C}$ is defined by $\mathscr{C} = \mathscr{C}_1 + \mathscr{C}^+ + \mathscr{C}^- - \Gamma$. The area enclosed by $\mathscr{C}$ is $\mathscr{A}$. Define a function $\tilde{q}(x_1, x_2)$ such that $\tilde{q} = 1$ on Γ and $\tilde{q} = 0$ on $\mathscr{C}_1$. In principle $\tilde{q}$ can be any function as long as it is sufficiently smooth. Recall from Eq. (3.47) that for a crack that grows straight ahead

$$G = J = \int_\Gamma \left(W n_1 - t_i \frac{\partial u_i}{\partial x_1} \right) d\Gamma$$

where Γ is a contour that starts and stops on the crack line and $\mathbf{n}$ is the outward unit normal to Γ. Replacing $\mathbf{n}$ with $-\mathbf{m}$ on Γ and noting that $\tilde{q} = 0$ on $\mathscr{C}_1$,

$$J = -\oint_\mathscr{C} \left(W m_1 - \sigma_{ij} m_j \frac{\partial u_i}{\partial x_1} \right) \tilde{q}\, d\mathscr{C} - \int_{\mathscr{C}^+ + \mathscr{C}^-} \sigma_{ij} m_j u_{i,1} \tilde{q}\, d\mathscr{C}.$$

The second integral equals zero since on the crack faces $\mathbf{m} = \pm m_2 \mathbf{e}_2$, thus $\sigma_{ij} m_j = \pm \sigma_{i2} = 0$. Thus J has been converted into an integral over a closed contour. The divergence theorem can now be used to convert J into an area integral:

$$\begin{aligned} J &= \int_\mathscr{A} [(\sigma_{ij} u_{i,1} \tilde{q})_{,j} - (W\tilde{q})_{,1}] dA, \\ &= \int_\mathscr{A} [\sigma_{ij} u_{i,j1} \tilde{q} + \sigma_{ij} u_{i,1} \tilde{q}_{,j} - W_{,1} \tilde{q} - W \tilde{q}_{,1}] dA \quad (\text{using} \sigma_{ij.j} = 0). \end{aligned}$$

Noting that $W_{,1} = \sigma_{ij} u_{i,j1}$

$$J = \int_\mathscr{A} [\sigma_{ij} u_{i,1} \tilde{q}_{,j} - W \tilde{q}_{,1}] dA. \tag{5.16}$$

Typically the function $\tilde{q}$ is expressed using the same shape functions that interpolate displacement, i.e. $\tilde{q} = \sum_{i=1}^{n} N_i \tilde{q}_i$, where N_i are the shape functions, $\tilde{q}_i$ are the nodal values of $\tilde{q}$ at nodes $i = 1, n$, see next section for further information on shape functions.

Thus J may be calculated as an area integral over any annular region surrounding the crack tip. The domain integral approach is generally very accurate even with modest mesh refinement since it does not rely on capturing the exact crack tip singular stress field, rather on correctly computing the strain energy in the region surrounding the crack tip.

5.4.5 Crack Tip Singular Elements

All of the above methods for computing stress intensity factors rely on the accurate calculation of the stress, displacement and energy fields. Since all of the methods use information from a small distance away from the crack tip they are somewhat forgiving of errors induced by not capturing the exact crack tip singular stress field. However, more accurate results could be obtained by capturing the crack tip singular stress field. Since we know that in elastic materials the crack tip stresses are singular as $1/\sqrt{r}$ this singularity can be built into the finite element calculation from the start. A number of methods to produce singular crack tip stresses have been developed, some of which require special elements and some of which can be used with standard elements. We will focus on 2D quarter-point elements that can be implemented using standard elements [21, 22].

In the finite element method the displacement field and the coordinates are interpolated using shape functions. Let $x = x_1,\ y = x_2,\ u = u_1,\ v = u_2$. Consider the 2D, 8 node isoparametric element shown in Fig. 5.9a. The parent element in the (ξ, η) space is mapped to an element in the physical space (x, y) using the shape functions, $N_i(\xi, \eta)$ and nodal coordinates (x_i, y_i),

$$\begin{aligned} x(\xi,\eta) &= \sum_{i=1}^{8} N_i(\xi,\eta)x_i, \\ y(\xi,\eta) &= \sum_{i=1}^{8} N_i(\xi,\eta)y_i. \end{aligned} \tag{5.17}$$

The displacements are interpolated in the same way, i.e. using the nodal displacements, (u_i, v_i)

$$\begin{aligned} u(\xi,\eta) &= \sum_{i=1}^{8} N_i(\xi,\eta)u_i, \\ v(\xi,\eta) &= \sum_{i=1}^{8} N_i(\xi,\eta)v_i. \end{aligned} \tag{5.18}$$

Fig. 5.9 Parent elements in (ξ, η) plane and mapped quarter-point elements. (**a**) 8 node rectangular element, (**b**) Triangular element formed by collapsing nodes 4,8,1 to a single point. With the collapsed node element the element edges must be straight in order to obtain accurate solutions [23]. (**c**) Natural triangular element. (Recommended element for linear problems)

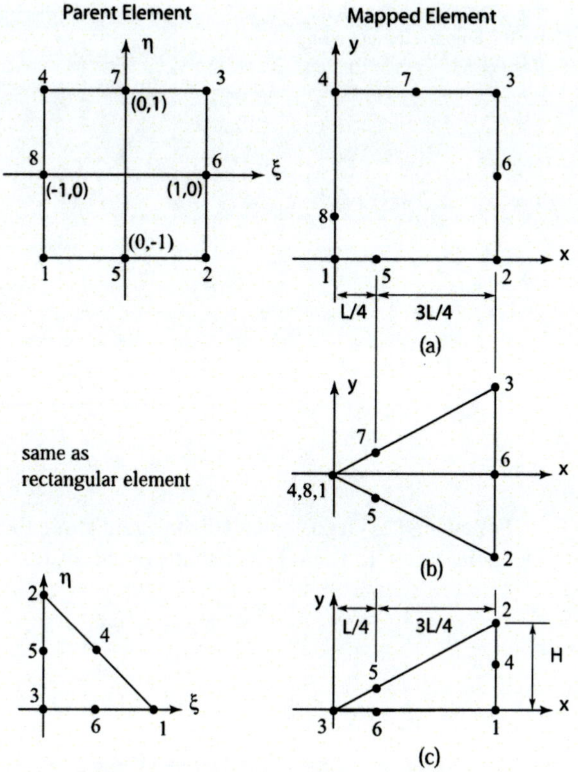

The shape functions have the property that $N_i = 1$ at node i and $N_i = 0$ at all other nodes. The shape functions for the 8 node element are

$$
\begin{aligned}
N_1 &= \frac{-(\xi-1)(\eta-1)(1+\eta+\xi)}{4},\\
N_2 &= \frac{(\xi+1)(\eta-1)(1+\eta-\xi)}{4},\\
N_3 &= \frac{(\xi+1)(\eta+1)(-1+\eta+\xi)}{4},\\
N_4 &= \frac{-(\xi-1)(\eta+1)(-1+\eta-\xi)}{4},\\
N_5 &= \frac{(1-\xi^2)(1-\eta)}{2},\\
N_6 &= \frac{(1-\eta^2)(1+\xi)}{2},\\
N_7 &= \frac{(1-\xi^2)(1+\eta)}{2},\\
N_8 &= \frac{(1-\eta^2)(1-\xi)}{2}.
\end{aligned}
\tag{5.19}
$$

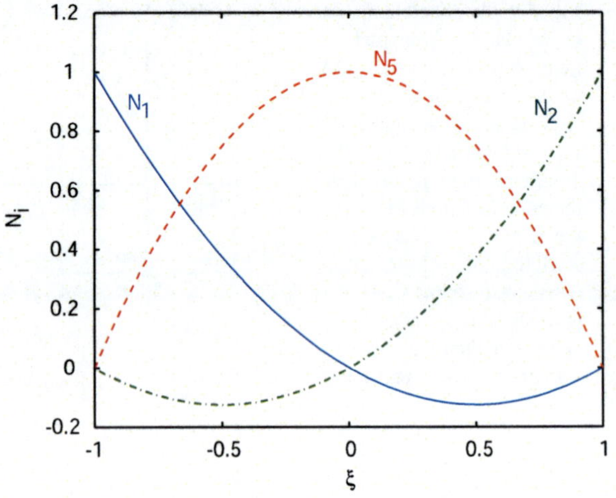

Fig. 5.10 Shape functions for the 8 node element along $\eta = -1$

Following [22] let us calculate the strain along the bottom of the 8 node element, i.e. along $\eta = -1$. The relevant shape functions on $\eta = -1$ are

$$\begin{aligned} N_1 &= -\frac{1}{2}\xi(1-\xi), \\ N_2 &= \frac{1}{2}\xi(1+\xi), \\ N_5 &= (1-\xi^2). \end{aligned} \tag{5.20}$$

These functions are plotted in Fig. 5.10.

The normal strain γ_{xx} is

$$\gamma_{xx} = \frac{\partial u}{\partial x} = \sum_{i=1,2,5} \frac{\partial N_i}{\partial \xi}\frac{\partial \xi}{\partial x} u_i, \tag{5.21}$$

where $\partial N_1/\partial\xi = \xi - 1/2$, $\partial N_2/\partial\xi = \xi + 1/2$, and $\partial N_5/\partial\xi = -2\xi$.

Consider first the case in which the node 5 (mid-side node) is located at the center of the element, so that $x_1 = 0,\ x_2 = L,\ x_5 = L/2$. Applying Eqs. (5.17) and (5.20),

$$\begin{aligned} x(\xi) &= \frac{\xi(1+\xi)}{2}L + (1-\xi^2)\frac{L}{2} \\ &= \frac{L}{2} + \frac{\xi L}{2}. \end{aligned}$$

Thus $\xi = (2x - L)/L$ and $\partial\xi/\partial x = 2/L$. Since none of the $\partial N_i/\partial\xi$ are singular, and since $\partial\xi/\partial x$ is non-singular, Eq. (5.21) results in a non-singular strain.

Now let us move the position of node 5 to the quarter-point, i.e. let $x_5 = L/4$, keeping $x_1 = 0$ and $x_2 = L$. Now

$$x(\xi) = \frac{\xi(1+\xi)}{2}L + (1-\xi^2)\frac{L}{4}$$

$$= \frac{L}{4}(\xi^2 + 2\xi + 1).$$

Solve for ξ and differentiate,

$$\xi(x) = -1 + 2\sqrt{\frac{x}{L}},$$

$$\frac{\partial \xi}{\partial x} = \frac{1}{\sqrt{xL}}.$$

Let $u_1 = 0$ then from Eq. (5.21), γ_{xx} along $\eta = 1$ is

$$\gamma_{xx} = \frac{1}{\sqrt{xL}}\left[-2\xi u_5 + \left(\frac{1}{2} + \xi\right)u_2\right],$$

$$= \frac{1}{\sqrt{xL}}\left[\frac{u_2}{2} + (u_2 - 2u_5)\xi\right].$$

Substituting $\xi = -1 + 2\sqrt{x/L}$ and simplifying,

$$\gamma_{xx} = \frac{4}{L}\left(\frac{u_2}{2} - u_5\right) + \frac{1}{\sqrt{xL}}\left(2u_5 - \frac{u_2}{2}\right). \tag{5.22}$$

Thus simply moving the mid-side nodes to the quarter points results in the desired $1/\sqrt{r}$ singularity. Other elements can also be used. The triangular element formed by collapsing the $\xi = -1$ side nodes to one point and moving the mid-side nodes to $x = L/4$ as shown in Fig. 5.9b also has $1/\sqrt{r}$ singularity [22].

The rectangular element has the drawback that it does not allow the crack to be surrounded by singular elements and hence to accurately capture the θ variation of stress, which might be needed for crack direction calculations. In addition, except along the element edges, the $1/\sqrt{r}$ singularity exists only in a small region near the crack tip [9]. The collapsed element has the drawback that the meshing will be somewhat more difficult to implement and that unless the element edges are straight, as shown in Fig. 5.9b, the $1/\sqrt{r}$ does not exist on straight lines coming from the crack tip and the accuracy in computing K will be degraded [23].

A better choice might be the natural triangle quarter point element shown in Fig. 5.9c [23]. Not only can the crack tip be easily surrounded by elements, the meshing is simple and the element gives the $1/\sqrt{r}$ on all lines coming from the crack tip. The shape functions for the 6 node triangular element are

$$N_1 = (2\xi - 1)\xi,$$

$$N_2 = (2\eta - 1)\eta,$$

$$N_3 = ((2(1 - \xi - \eta) - 1)(1 - \xi - \eta),$$

$$N_4 = 4\xi\eta,$$

$$N_5 = 4\eta(1 - \xi - \eta),$$

$$N_6 = 4\xi(1-\xi-\eta).$$

Consider the edge along the x axis, $\eta = 0$ in the parent element. With $x_6 = L/4$ and $x_1 = L$,

$$\begin{aligned} x &= x_1 N_1 + x_6 N_6 \\ &= L(2\xi - 1)\xi + \frac{L}{4} 4\xi(1-\xi) \\ &= L\xi^2. \quad \text{Thus} \\ \xi &= \sqrt{\frac{x}{L}}. \end{aligned}$$

The u displacement along this line is

$$\begin{aligned} u &= u_1 N_1(\xi, 0) + u_6 N_6(\xi, 0) + u_3 N_3(\xi, 0) \\ &= u_1(2\xi - 1)\xi + u_3((2(1-\xi) - 1)(1-\xi) + u_6 4\xi(1-\xi). \end{aligned}$$

Substituting $\xi = \sqrt{\frac{x}{L}}$ and simplifying

$$u = u_3 + \sqrt{\frac{x}{L}}[4u_6 - 3u_3 - u_1] + \frac{x}{L}[2u_1 + 2u_3 - 4u_6], \tag{5.23}$$

$$\gamma_{xx} = \frac{\partial u}{\partial x} = \frac{1}{\sqrt{xL}}[-\frac{u_1}{2} - \frac{3u_3}{2} + 2u_6] + \frac{1}{L}[2u_1 + 2u_3 - 4u_6]. \tag{5.24}$$

Thus the natural triangle quarter-point element has constant and $1/\sqrt{r}$ strain terms, reproducing the first two terms of the Williams crack tip solution. Note that the full strain field can be shown to have $1/\sqrt{r}$ singularity, see [22].

With any of these elements accuracy per unit computational time should be significantly better than with the use of non singular elements. For example, Banks-Sills and Sherman [24] compared displacement extrapolation, J integral and total energy approaches using singular and non-singular elements. For a central cracked plate under tension using 100 8 node elements displacement extrapolation had an error in stress intensity factor of 5.4% using regular elements and 1.8% using quarter point elements. For the same problem, using the total energy method with a mesh of 121 8 node elements the error was 2.4% using regular elements and 0.37% using quarter point elements.

5.4.6 Example Calculations

Consider a plane stress SENT sample with properties, loading and dimensions (in units of N and mm) of $E = 1000$, $\nu = 0.25$, $\sigma = 1$, $a = 20$, $b = 50$, $h = 100$.

Table 5.2 Summary of results of FEM calculation using displacement correlation and domain integral methods for mesh with singular (1/4 point) elements. All quantities in N and mm units

Mesh size	K_I from disp. corr	Error, %	J	K_I from J	Error %
0.25a	16.74	0.4	0.2796	16.72	0.3
0.125a	16.79	0.7	0.2798	16.73	0.4
0.0625a	16.76	0.5	0.2800	16.73	0.4

The mode I stress intensity factor will be computed using displacement correlation, global energy, domain integral and MCCI methods. From [4] the theoretical results are $K_I = 16.67$ N/mm$^{3/2}$ and $J = 0.2778$ N/mm.

5.4.6.1 Displacement Correlation and Domain Integral with 1/4 Point Elements

The SENT sample was modeled in plane stress with a unit thickness using meshes focused on the crack tip as shown in Fig. 5.11. Quadratic elements were used for each mesh and to capture the $1/\sqrt{r}$ stress singularity the center nodes of the natural triangle crack tip nodes were moved to the 1/4 point as shown in the figure. The mesh refinement is characterized by the radial length of the elements at the crack tip. The coarse, medium and fine meshes had element lengths of 5.0, 2.5, and 1.25 mm, or $0.25a$, $0.125a$, and $0.0625a$, where a is the crack length.

The computed crack opening displacements and the K_I^* values calculated for the displacement correlation method are shown in Figs. 5.12(a) and 5.12(b). Note that K_I^* vs. r is nearly linear. K_I^* was extrapolated back to the crack tip by using a linear fit to $K_I^*(r)$ and noting the y intercept value of the fit. The computational results for K_I and J are summarized in Table 5.2. Note that the theoretical results are accurate to 0.5%, thus the numerical and theoretical results agree to within the accuracy of the theory. (It is likely that the correct value of K_I is 16.73, the value given by the domain integral method with the finest mesh.) In any case, it is seen that both methods provide sufficient accuracy even with the coarsest mesh. However, in every case the domain integral result is more accurate. Accurate results with the domain integral method could be achieved with even coarser meshes.

5.4.6.2 Global Energy

Using the medium mesh shown in Fig. 5.11 the total elastic strain energy was computed for two cases, one with crack length of $a = 20$ mm and another for crack length of $a = 20.1$ mm. The total energies in each case were 3.30598 N·mm/mm and 3.31962 N·mm/mm respectively. Note that a symmetry model was used, so the energies must be multiplied by two. Thus the energy release rate was estimated to be $G \approx 2 \times (3.31962 - 3.30598)/(0.1) = 0.2728$ N/mm or $K_I = \sqrt{GE} =$ 16.52 N/mm$^{3/2}$, an error of 0.8%.

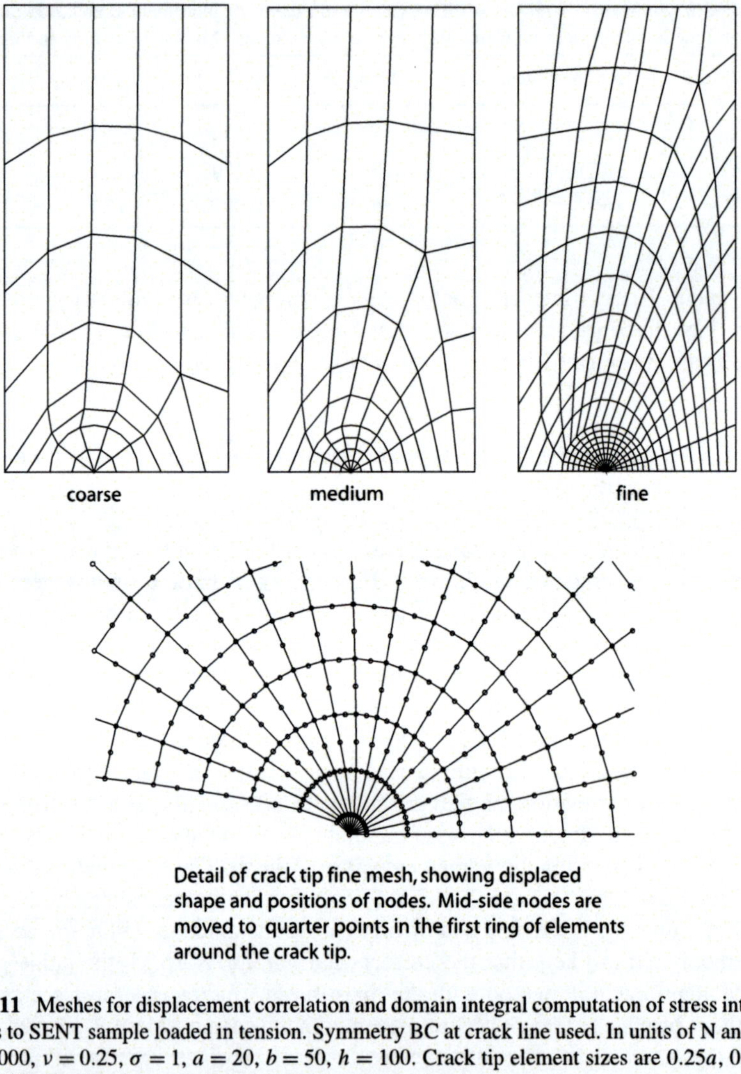

Fig. 5.11 Meshes for displacement correlation and domain integral computation of stress intensity factors to SENT sample loaded in tension. Symmetry BC at crack line used. In units of N and mm, $E = 1000$, $\nu = 0.25$, $\sigma = 1$, $a = 20$, $b = 50$, $h = 100$. Crack tip element sizes are $0.25a$, $0.125a$, and $0.062a$ for the coarse, medium and fine meshes respectively

5.4.6.3 Modified Crack Closure Integral

The Modified Crack Closure Integral or MCCI, method is most readily used with rectangular crack tip elements of even size, without singular elements. A medium and a fine mesh were used, Fig. 5.13. The results are summarized in Table 5.3. The fine mesh 4 node model and the medium mesh 8 node model have the same number of degrees of freedom, yet K_I calculated using the second order method with the 8 node elements in the medium mesh is much more accurate than that calculated using the fine mesh with 4 node elements. The results show that reasonably good accuracy can be obtained with the MCCI method even for meshes of modest refinement.

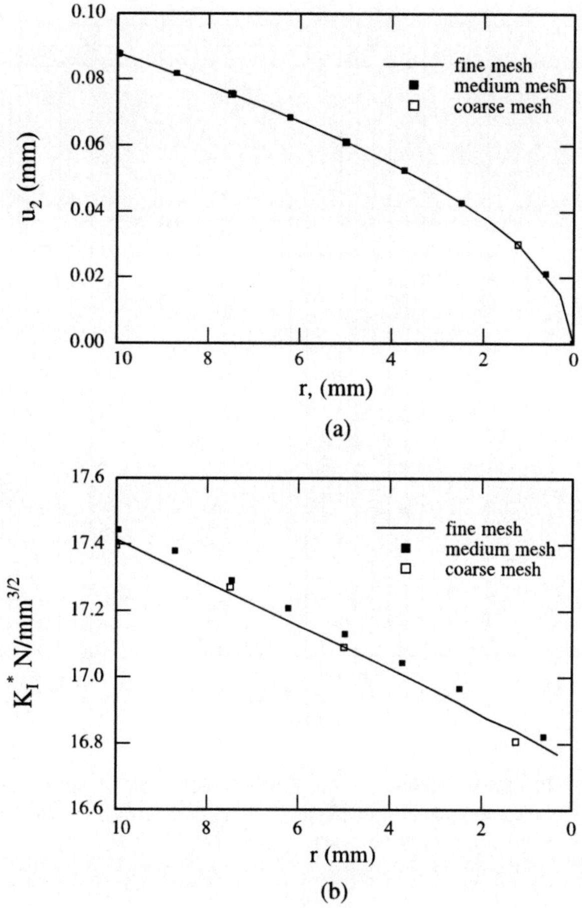

Fig. 5.12 Results from FEM analysis of SENT specimen with quarter-point elements. (**a**) Crack line displacement. (**b**) K_I^* for extrapolation of stress intensity factor in displacement correlation method

Table 5.3 Summary of MCCI method computations. Results in N and mm units

Mesh size	G	K_I	Error, %
4 node elements			
0.125a	0.2569	16.03	−3.8
0.0625a	0.2687	16.39	−1.7
8 node elements			
0.125a	0.2694	16.41	−0.6
0.0625a	0.2748	16.58	−0.5

5.5 Experimental Methods

In some cases it may not be practical to determine stress intensity factors from analytical or computational methods. For example, perhaps the loading is not known, or is dynamic, or information about parts of the structure that would be needed for

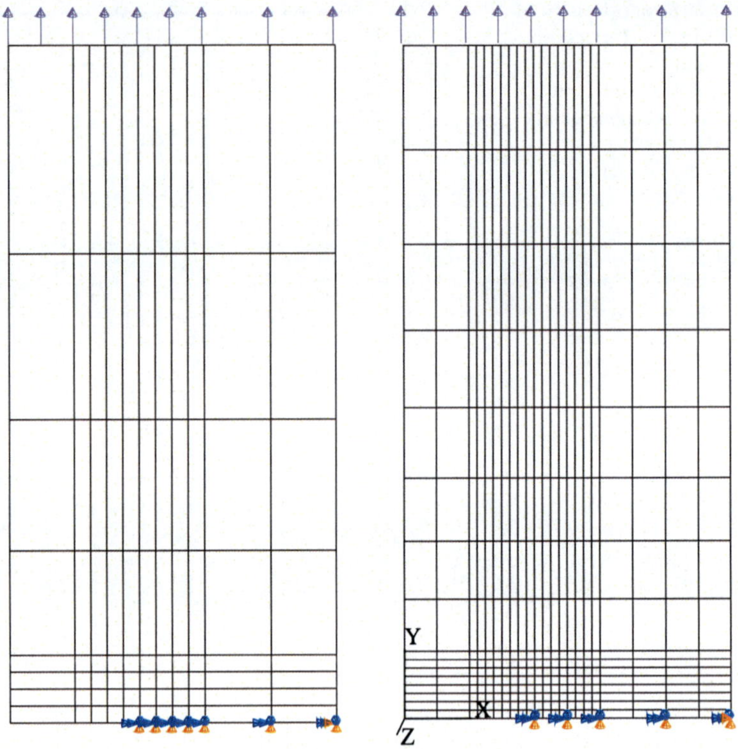

Fig. 5.13 Medium and fine meshes used for MCCI sample computation of K_I in SENT sample. Linear (4 node) and quadratic (8 node) elements were used. Medium mesh: crack tip element size $= 0.125a$. Fine mesh: crack tip element size $= 0.0625a$

a FEM analysis are missing. In such cases one may wish to determine the stress intensity factor experimentally, based on local measurements of stress, strain and displacement.

A number of experimental methods such as photoelasticity, the shadow spot (caustics) method, interferometry, digital image correlation, thermoelasticity and strain gauges have been developed for fracture research and could in principle be applied to determine stress intensity factors in a component. Digital image correlation, photoelasticity, strain gauges and thermoelasticity are discussed here. A number of other methods are described in [25].

5.5.1 Strain Gauge Method

Most engineers and technicians are familiar with resistance strain gauges. Furthermore strain gauges can be bonded to many materials and are relatively inexpensive as is the signal conditioning and other equipment needed. Thus the strain gauge

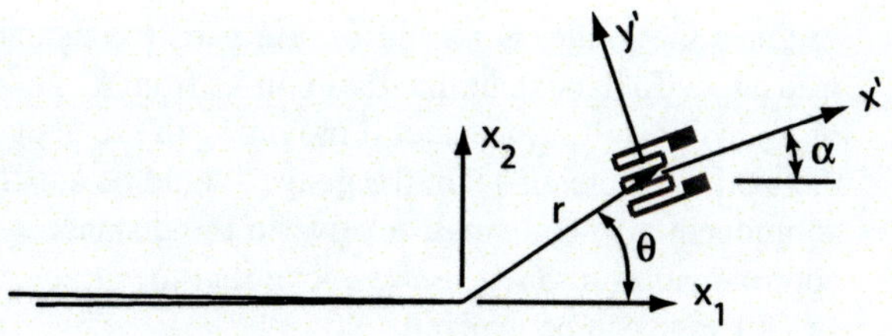

Fig. 5.14 Location of single strain gauge near crack tip for measurement of K_I. By selecting θ and α the strain in the x' direction will depend only on K_I for a three term expansion of the elastic, plane-stress crack tip fields

method [25, 26] is promising for experimentally determining stress intensity factor in real-world applications.

The strain gauge method assumes the presence of an elastic, plane-stress K field with higher order terms. Let us assume that a strain gauge, or strain gauges can be placed in a region near the crack tip where the stresses are accurately described by the first three terms of the Williams expansion, Eq. (2.49).

Consider placing a gauge at position (r, θ) with respect to the crack tip and with orientation α with respect to the x_1 line as shown in Fig. 5.14. Calculate the strain at this point and then rotate the strain components into the x', y' system. Dally and Sanford [26] show that the strain parallel to the gauge, i.e. in the x' direction, is given by

$$\begin{aligned} 2\mu\gamma_{x'x'} = {} & K_I(2\pi r)^{-1/2}\big[(1-\nu)/(1+\nu)\cos\theta/2 \\ & - (1/2)\sin\theta\sin(3\theta/2)\cos 2\alpha + (1/2)\sin\theta\cos(3\theta/2)\sin 2\alpha\big] \\ & + A_0\big[(1-\nu)/(1+\nu) + \cos 2\alpha\big] \\ & + A_{1/2}\big[(1-\nu)/(1+\nu) - \sin^2(\theta/2)\cos 2\alpha + (1/2)\sin\theta\sin 2\alpha\big]. \end{aligned} \tag{5.25}$$

There are three parameters, but only one measurement. What do you do? Fortunately the A_0 term can be eliminated by choosing $\alpha = \alpha^*$ such that

$$\cos 2\alpha^* = -\frac{1-\nu}{1+\nu}. \tag{5.26}$$

The $A_{1/2}$ term can be eliminated if the gauge is located so that $\theta = \theta^*$, where

$$\tan\frac{\theta^*}{2} = -\cot 2\alpha^*. \tag{5.27}$$

Now K_I can be determined from the strain gauge

$$K_I = \frac{2\mu\gamma_{x'x'}}{\sqrt{2\pi r}} f(\nu), \tag{5.28}$$

where

$$\begin{aligned} f(\nu) = {} & \big((1-\nu)/(1+\nu)\cos\theta^*/2 - (1/2)\sin\theta^*\sin(3\theta^*/2)\cos 2\alpha^* \\ & + (1/2)\sin\theta^*\cos(3\theta^*/2)\sin 2\alpha^*\big). \end{aligned} \tag{5.29}$$

Once the gauge is placed on the part, the actual value of θ and α should be measured. To the extent that these differ from θ^* and α^* errors in the determination of K_I will result. Thus careful mounting of the gauges is critical to the accuracy of the method. Note also that the gauge should be small relative to the crack length or component size and small relative to the distance r from the crack tip so that the approximation implicit in the above that the strain is constant over the area of the strain gauge can be realized.

5.5.2 Photoelasticity

The strain gauge method above uses data from only a single point around the crack tip and also assumes that the stress field can be expressed using just 3 terms in the series expansion. Full field methods such as photoelasticity allow one to use data from the entire region surrounding the crack and remove the need to make restrictive assumptions about the nature of the crack tip fields. A drawback of the photoelastic method is that it requires either the use of photoelastic models built of a birefringent polymer or the placing of photoelastic coatings on the component.

Photoelasticity is an optical method for the visualization and measurement of in-plane stresses. It can be used either with models built from a birefringent polymer or on actual parts that have been coated with a birefringent material [27]. A simulated photoelastic pattern for a tensile crack in a plate is shown in Fig. 5.15. The fringes are contours of constant difference in principal stress, $\sigma_1 - \sigma_2$, which using the generalized Westergaard solution for crack tip fields can be expressed as [28]

$$\left(\frac{Nf_\sigma}{2B}\right)^2 = F_1^2 + F_2^2 \tag{5.30}$$

where B is the plate thickness, f_σ is the stress-optic coefficient, N is the fringe order (or number) and

$$\begin{aligned} F_1 &= \sum_{n=0}^{\infty}\left(n-\frac{1}{2}\right)A_n r^{n-1/2}\sin\theta\sin(n-3/2)\theta \\ &\quad + \sum_{m=0}^{\infty} B_m r^m[m\sin\theta\sin(m\theta)+\cos(m\theta)] \\ F_2 &= -\sum_{n=0}^{\infty}\left(n-\frac{1}{2}\right)A_n r^{n-1/2}\sin\theta\cos(n-3/2)\theta \\ &\quad - \sum_{m=0}^{\infty} B_m r^m[m\sin\theta\cos(m\theta)+\sin(m\theta)]. \end{aligned} \tag{5.31}$$

The coefficient A_0 corresponds to the stress intensity factor, $K_I = \sqrt{2\pi}A_0$. The method for finding K_I is to select a number of points (r_i, θ_i), $i = 1, k$ surrounding

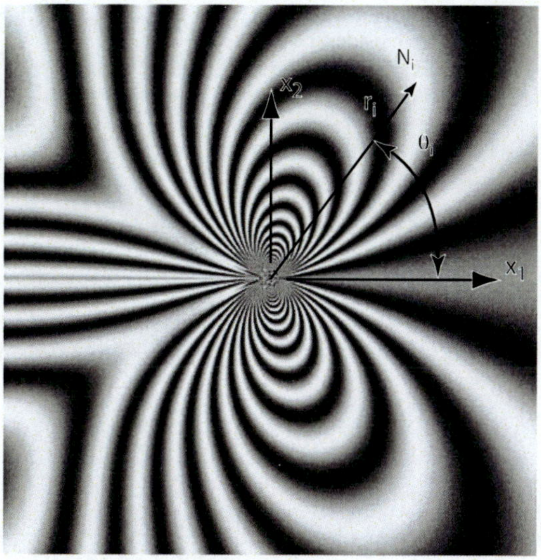

Fig. 5.15 Simulated photoelastic fringes for a tensile crack. The fringe order N_i is determined at an array of points (r_i, θ_i)

the crack and to determine the fringe order N_i at each of these k points. Truncating the series for F_1 and F_2 at $n = nmax$ and $m = mmax$ you will have k equations

$$\left(\frac{N_i f_\sigma}{2b}\right)^2 = F_1^2(r_i, \theta_i) + F_2^2(r_i, \theta_i), \quad i = 1, k, \tag{5.32}$$

for $nmax + nmax$ unknowns, $\{A_0, \ldots, A_{nmax}, B_0, \ldots, B_{mmax}\}$. These expressions are nonlinear in the unknowns. An iterative scheme for the solution of these equations is outlined in [28]. Other nonlinear least squares methods can also be employed. Convergence can be verified by increasing the order of the series solutions and by increasing the number of data points until K_I converges.

5.5.3 Digital Image Correlation

Digital image correlation (DIC) is a full-field method for the measurement of two- and three-dimensional displacements on the surface of an object under deformation. The measured displacement field can be used to determine the stress intensity factor, [29], crack tip opening displacement (CTOD) [30] or other fracture parameters.

In 2D, the method works by correlating images obtained before and after deformation of surfaces covered with a speckle pattern. A simulated pair of images of the underformed and deformed surfaces of a plate containing a tensile crack is shown in Fig. 5.16. Point P at (x, y) moves to a new position, p, at $(x + u, y + v)$ where u,v are the x,y displacement components. Points Q_i in the neighborhood of P, at initial positions $(x + dx_i, y + dy_i)$ move to q_i with positions $(x + u + (1 + u_{,x})dx_i + u_{,y}\, dy_i, y + v + v_{,x}\, dx_i + (1 + v_{,y})dy_i)$. Assuming locally homogeneous deformation, $\{u, v, u_{,x}, u_{,y}, v_{,x}, v_{,y}\}$ are uniform over a small

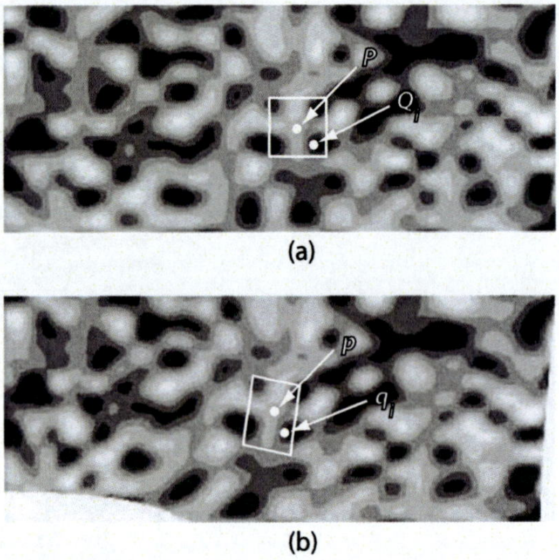

Fig. 5.16 Simulated speckle patterns around a tensile crack. (**a**) Undeformed. (**b**) Deformed

region surrounding p. Let $I(x, y)$ and $I'(x, y)$ be the measured intensity pattern of the deformed and undeformed images respectively. Let $\tilde{I}(x, y)$ and $\tilde{I}'(x, y)$ be continuous interpolations of the discretely sampled I and I'. For a given point P the displacement is found by maximizing the normalized cross correlation coefficient, [30]

$$C = \left(\sum_i \tilde{I}(x + dx_i, y + dy_i)\tilde{I}'(x + u + (1 + u_{,x})dx_i + u_{,y}\,dy_i, y + v + v_{,x}\,dx_i + (1 + v_{,y})dy_i)\right) \Big/ \left(\left[\sum_i \tilde{I}^2(x + dx_i, y + dy_i)\right]^{1/2} \left[\sum_i \tilde{I}'^2(x + u + (1 + u_{,x})dx_i + u_{,y}\,dy_i, y + v + v_{,x}\,dx_i + (1 + v_{,y})dy_i)\right]^{1/2}\right) \tag{5.33}$$

with respect to the deformation parameters $\{u, v, u_{,x}, u_{,y}, v_{,x}, v_{,y}\}$, where the summation is taken over all Q_i in the neighborhood of P. The term $\tilde{I}'(x + u + (1 + u_{,x})dx_i + u_{,y}\,dy_i, y + v + v_{,x}\,dx_i + (1 + v_{,y})dy_i)$ represents the intensity of the deformed image shifted by $(u + (1 + u_{,x})dx_i + u_{,y}\,dy_i, v + v_{,x}\,dx_i + (1 + v_{,y})dy_i)$. Note that other correlations can be taken as well, see [30]. Remarkably, this method can resolve displacements to approximately ± 0.02 pixels. This process is repeated over many points, P, to provide the u and v displacements over a grid of points. K_I can be determined by linear least squares fitting of the measured crack tip displacements to the crack tip displacement field, taking a multi-term expansion of the crack tip displacement field, Eq. (2.50).

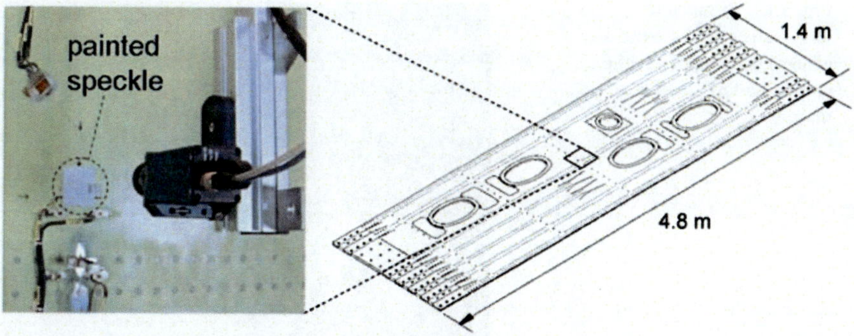

Fig. 5.17 Digital image correlation applied to measure stress intensity factors in a large stiffened skin aircraft panel. *Left image* shows the speckle pattern painted onto a 50 × 60 mm region at the crack tip. *Right image* shows the panel and the region studied. Courtesy of Prof. E. Patterson, from [31]

Figure 5.17 shows a sample application of DIC to an engineering structure. In this example a small fatigue crack is introduced in a very large, stiffened skin, aluminum alloy panel, representative of the wing structure of a large transport aircraft [31]. A speckle pattern was painted onto the structure in a small area around the crack tip. As the panel was monotonically loaded the crack grew slowly and a succession of digital images was obtained. The stress intensity factors K_I and K_{II} were determined following the procedures outlined above. The results show the crack growing for about 10 mm in Mode-I with an increasing K_I. When K_I reached about 40 MPa$\sqrt{\text{m}}$ the crack kinked and grew with under mixed, Mode-I and Mode-II conditions. More important than the specific results, this work demonstrates the capability to apply DIC for fracture problems in large scale structures.

5.5.4 Thermoelastic Method

Another full field, non-contact method for measuring stress intensity factors is thermoelasticity, which exploits the small temperature changes associated with elastic straining under adiabatic conditions. As gasses heat or cool as they are compressed or expanded, so do solids. Under adiabatic conditions, i.e. no heat flow, the change in temperature in an elastic deformation is proportional to the change in dilatation, or in terms of stress, to the change in the first invariant of the stress tensor,

$$\Delta T = \frac{-\alpha T}{\rho c} \Delta(\sigma_{ii}), \tag{5.34}$$

where ΔT is the temperature change, α is the coefficient of thermal expansion, ρ is the density and c is the specific heat. A simulation of the temperature field around a mode-I crack under cyclic, elastic loading is shown in Fig. 5.18. In practice the thermoelastic method is used by measuring the temperature change from the front

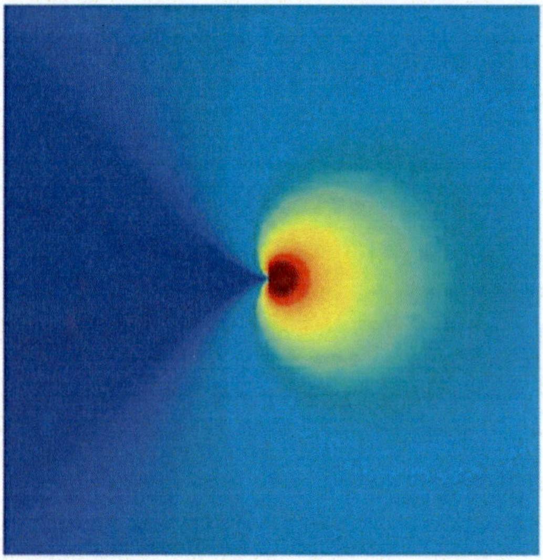

Fig. 5.18 Simulated temperature field around a Mode-I crack under cyclic, elastic loading. Arbitrary units

of a black painted surface using an infrared camera. The load is applied cyclically at a frequency high enough, e.g. 10 Hz, that the deformation can be considered adiabatic. The system must be calibrated to relate the measured IR intensity, I, to a known $\Delta\sigma_{ii}$. Since $\sigma_{33} = 0$ on the free surface, the sum of the change of the in-plane principal stresses is related to the measured IR intensity by

$$\Delta(\sigma_{11} + \sigma_{22}) = AI, \tag{5.35}$$

where A is an experimentally determined calibration constant [32].

A number of methods can be used to extract the cyclic stress intensity factor, ΔK_I from the IR intensity, I [33]. Note that in a tensile crack, if the K_I field dominates, then from Eq. (2.49), ΔK_I is related for the IR intensity by

$$\frac{2\Delta K_I}{\sqrt{2\pi r}} \cos\frac{\theta}{2} = AI(r,\theta), \tag{5.36}$$

or

$$\Delta K_I = \frac{\sqrt{2\pi}A}{2\cos\theta/2}\left(\sqrt{r}I(r,\theta)\right). \tag{5.37}$$

In an approach analogous to displacement and stress correlation in FEM, ΔK_I can be found by fitting the measured $I(r,\theta)$ along a radial line extending from the crack to Eq. (5.37) above.

However, such an approach throws away a great deal of useful information and only addresses Mode-I cracks. An alternative approach is to represent the crack tip fields using multiple terms for the crack tip stress fields, Eqs. (2.49) and (2.61). From the thermal field determine $\sigma_{11} + \sigma_{22}$ at a grid of points surrounding the crack tip, then using least squares, fit the coefficients of the crack tip stress field to the measured data. Since plastic deformation produces additional temperature changes care must be taken in the thermoelastic method to include only IR data taken from outside the crack tip plastic zone. As will be shown in Chap. 7, crack tip plastic deformation can have the effect of shifting the elastic field outside the plastic zone to a location ahead of the actual crack tip. Thus an additional parameter that can be used to increase accuracy is to fit the crack tip location, see [33] and [32] for further details.

5.6 Exercises

1. Calculate the stress intensity factor for a semi-infinite crack with point loads $\pm P_2 \mathbf{e}_2$ applied at a distance ℓ behind the crack tip.
2. Show that for the case of uniform p_2, the stress intensity factor from Eq. (5.1) reduces to Eq. (2.80).
3. Calculate the stress intensity factor for a finite crack subjected to uniform tractions over a small region near both crack tips, i.e. $p_2 = \sigma_0$ on $(a - \ell \leq |x| \leq a)$ and $p_2 = 0$ on $(|x| < a - \ell)$.
4. When a body has more than one crack the stress fields from each crack will affect the other. However, it is reasonable as well to expect that when the cracks are widely spaced that their interaction will be minimal. To estimate the distance beyond which cracks have little interaction, calculate the minimum value of b/a for which K_I for the array of cracks shown in Fig. 5.1 differs by less than 5% from that of a single crack.
5. Calculate the energy release rate for the 4 point bending specimen, Fig. 5.2 using the J-integral method.
6. Extend the calculation of energy release rate for the 4 point bending specimen, Fig. 5.2 to the case when the two beams have different elastic moduli.
7. Calculate the energy release rate for the sample shown in Fig. 5.2 replacing the bending moment M with a tensile load applied at the centroid of the *uncracked* section.
8. Calculate K_I and K_{II} for both the left and the right crack tips for line loads P_2 and P_1 located distance b off-center from a finite crack of length $2a$.
9. For the strain gauge method, determine the angles α^*, θ^* and the numerical value of the function $f(\nu)$ as a function of ν. Verify that for $\nu = 0.30$, $\alpha^* = 61.29°$ and $\theta^* = 65.19°$.
10. For the anti-plane shear crack in a circular cylinder, Fig. 5.4 we know from dimensional analysis that

$$K_{III} = \eta \tau_a \sqrt{R}.$$

   (a) Use boundary collocation to determine the numerical factor η for the following tractions $t_3 = \tau_a F(\theta)$ on the outer boundary of the circular cylinder:

$$t_3 = \sigma_a F(\theta) = \begin{pmatrix} \text{sign}(\theta) \\ \sin(\theta) \\ \sin(\theta/2) \end{pmatrix}. \tag{5.38}$$

(b) What do the load distributions $F(\theta) = \sin(\theta)$ and $F(\theta) = \sin(\theta/2)$ correspond to physically?

(c) For the case $F(\theta) = \text{sign}\theta$ examine the convergence of η as the number of collocation points and the number of terms in the stress field expansion are increased.

(d) Examine the coefficients A_n in your solutions to the above problems. In which of the load cases are the higher order terms most important? In which case are the higher order terms equal to zero? Why?

Note: To check your results you might want to plot σ_{32} along $\theta = \pm\pi$ and t_3 along $r = R$ to be sure that the crack surfaces are traction free and that the calculated tractions match the prescribed loads.

References

1. J.R. Rice, in *Fracture an Advanced Treatise*, vol. 2, ed. by H. Liebowitz (Academic Press, New York, 1968), pp. 191–311, chap. 3
2. K.B. Broberg, *Cracks and Fracture* (Academic Press, San Diego, 1999)
3. P.G. Charalambides, J. Lund, A.G. Evans, R.M. McMeeking, J. Appl. Mech. **56**, 77 (1988)
4. H. Tada, P.C. Paris, G.R. Irwin, *The Stress Analysis of Cracks Handbook* (ASME Press, New York, 2000)
5. Y. Murakami, S. Aoki, *Stress Intensity Factors Handbook* (Pergamon, Oxford, 1987)
6. Nascrac software. http://www.exponent.com/about/nascrac.html
7. NASGRO 4.0 User's Manual (SouthWest Research Institute, http://www.nasgro.swri.org/)
8. ASTM, *ASTM E 1820: Standard Test Method for Measurement of Fracture Toughness* (ASTM International, 2005)
9. A.R. Ingraffea, P.A. Wawrzynek, in *Comprehensive Structural Integrity*, vol. 3, ed. by R. de Borst, H. Mang (Elsevier, Oxford, 2003), Chap. 1
10. A.R. Ingraffea, in *Encyclopedia of Computational Mechanics*, vol. 2, ed. by E. Stein, R.D. Borst, T.J. Hughes (Wiley, New York, 2004), Chap. 11
11. S. Chan, I. Tuba, W. Wilson, Eng. Fract. Mech. **2**, 1 (1970)
12. H. Ansell, Bulging of cracked pressurized aircraft structures. Ph.D. thesis, Linköping University, Linköping, Sweden (1988). Linköping University Report No. LIU–TEK–LIC1988:11
13. H. de Lorenzi, Int. J. Fract. **19**, 183 (1982)
14. E. Rybicki, M. Kanninen, Eng. Fract. Mech. **9**, 931 (1977)
15. R. Sethuraman, S. Maiti, Eng. Fract. Mech. **30**, 227 (1988)
16. I.S. Raju, Eng. Fract. Mech. **28**, 251 (1987)
17. T. Ramamurthy, T. Krishnamurthy, K. Narayana, K. Vijayakumar, B. Dattaguru, Mech. Res. Commun. **13**, 179 (1986)
18. C.F. Shih, B. Moran, T. Nakamura, Int. J. Fract. **30**, 79 (1986)
19. F.Z. Li, C.F. Shih, A. Needleman, Eng. Fract. Mech. **21**, 405 (1985)
20. B. Moran, C.F. Shih, Int. J. Fract. **35**, 295 (1987)
21. R.D. Henshell, K.G. Shaw, Int. J. Numer. Methods Eng. **9**, 495 (1975)
22. R.S. Barsoum, Int. J. Numer. Methods Eng. **10**, 25 (1976)
23. C.E. Freese, D.M. Tracey, Int. J. Fract. **12**, R767 (1976)
24. L. Banks-Sills, D. Sherman, Int. J. Fract. **32**, 127 (1986)
25. J.S. Epstein (ed.), *Experimental Techniques in Fracture* (VCH, New York, 1993)
26. J.W. Dally, R.J. Sanford, Exp. Mech. **27**, 381 (1988)
27. A. Kobayashi, *Handbook on Experimental Mechanics* (Prentice Hall, New York, 1987)
28. R.J. Sanford, *Principles of Fracture Mechanics* (Prentice Hall, New York, 2003)
29. S.R. McNeil, W.H. Peters, M.A. Sutton, Eng. Fract. Mech. **28**, 101 (1987)

30. M.A. Sutton, S.R. McNeil, J.D. Helm, Y.J. Chao, *Photomechanics*. Topics in Applied Physics, vol. 77 (Springer, Berlin, 2000), Chap. Advances in Two-Dimensional and Three-Dimensional Computer Vision, pp. 323–372
31. Y. Du, F.A. Diaz-Garrido, R.L. Burguete, E.A. Patterson, in *Proceedings of the 2009 SEM Annual Conference and Exposition on Experimental and Applied Mechanics* (Society for Experimental Mechanics, 2009)
32. R.A. Tomlinson, A.D. Nurse, E.A. Patterson, Fatigue Fract. Eng. Mater. Struct. **20**(2), 217 (1997)
33. R.A. Tomlinson, E.J. Olden, Strain **35**(2), 49 (1999)

Chapter 6
Fracture Toughness Tests

Abstract To apply the theory developed in the previous chapters, fracture toughness values must be measured. A wealth of methods for the determination of critical values of the stress intensity factor, R-curve, energy release rate have been developed. A number of these fracture toughness test methods are described here. The choice of method to use will depend on the material being tested and on your goals. Methods that work for steel will not work for composite materials. If you need data for a critical application you'll want to use a standard, accepted test method. If you are developing a new material and want to determine toughness as part of the development or screening process you might choose another method.

6.1 Introduction

Until the day in which complex materials can be accurately simulated from the atomic level up, all of the above criteria require physical testing in order to determine the proper values and to test the validity of the criteria. As such a day is likely to be long or perhaps infinite in coming, some acquaintance with physical testing is required to understand and to apply fracture mechanics. A complete description of experimental methods in fracture would require several long books. Thus in this section the bare outlines of the equipment, measurements, basic test and sample types, standards and interpretation of data will be described, focusing on the fracture of materials for which linear elastic fracture mechanics is a good approximation. Both ASTM standard and several non-standard but useful methods will be outlined here. These outlines should not be used to substitute for a careful reading of the relevant ASTM standards or in-depth descriptions given in the references.

Physical tests related to elastic-plastic fracture will be briefly discussed in a separate chapter. The discussion here will focus on testing at ambient temperatures and in laboratory air environments. Although the principles remain essentially the same, elevated and low temperature tests require special equipment and considerations. The environment, for example the presence of corrosive agents, or of hydrogen, or water has a significant impact on crack propagation. Experimental methods for environmentally assisted and high temperature fracture deserve an entire book themselves.

A.T. Zehnder, *Fracture Mechanics*,
Lecture Notes in Applied and Computational Mechanics 62,
DOI 10.1007/978-94-007-2595-9_6,

6.2 ASTM Standard Fracture Test

To apply fracture mechanics to the prediction of failure in real-world applications one needs to know the fracture toughness of the material at hand. To ensure that the toughness values used in the application are correct, great care must be taken in the tests used to determine the toughness. ASTM E399, "Standard test method for plane-strain fracture toughness of metallic materials," [1] provides a step by step method for measuring fracture toughness and for ensuring that the test data are valid. The method is based on ensuring that sufficient constraint exists to provide plane-strain conditions at the crack. Since it has been observed that as test samples get thinner and the plane-strain constraint is relaxed the measured fracture toughness increases. Hence by ensuring plane-strain conditions E399 provides a lower bound to the fracture toughness, a conservative approach, since the actual toughness may be higher if the data are applied to a thinner section. Fracture toughness values determined by a test that satisfies all the E399 criteria are denoted by K_{IC}. To avoid confusion this notation should be reserved for toughness values determined through E399. The E399 test is reviewed here. However, this review is not a substitute for the standard. If you want to determine K_{IC} you must read and then carefully follow each step of the test standard. A similar, but separate standard is available for aluminum alloys [2].

6.2.1 Test Samples

Fracture toughness may depend on the orientation of the fracture surface and direction of crack growth. This is particularly true in crystalline materials in which certain planes may have lower fracture energies than others, or in composite materials where the geometry of the reinforcing fibers or particles generally results in elastic anisotropy and in great differences in fracture toughness in different directions. Toughness may also vary with direction in polycrystalline metals due to the texture induced in grains by rolling or extrusion. To standardize the reporting of toughness data a lettering scheme has been introduced in which specimens are referred to as having a T-L, or L-T or other orientation. The method is as follows. As shown in Fig. 6.1 the rolling or extrusion direction is referred to as L, the transverse direction as T and the thickness as S. In the orientation designation the first letter refers to the direction normal to the crack plane while the second refers to the direction or prospective crack growth. Only two possible orientations are shown in the figure. Many more are possible, see [1] for complete details.

Four geometries of test specimens may be used, the compact tension (CT), single edge notch bend (SENB), disk shaped CT and arc-shaped specimen. The CT and SENB geometries are given in Table 5.1 along with the required dimensions and formulas for stress intensity factor, crack mouth opening displacement and load line

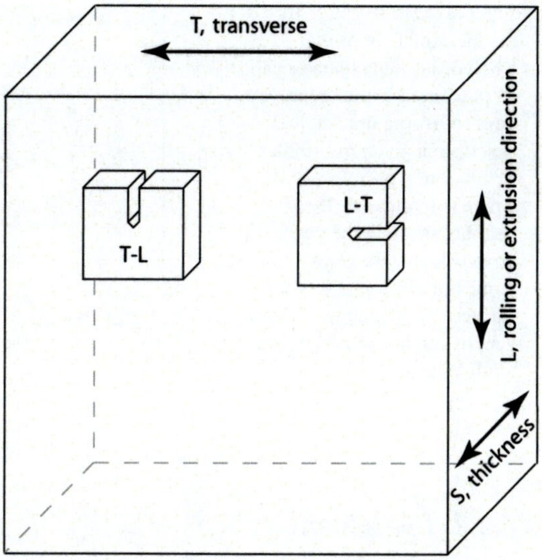

Fig. 6.1 Labeling of fracture test specimen orientations. First letter refers to direction normal to crack plane. Second letter refers to prospective crack growth direction. For complete list of orientations and their labeling see [1]

displacement (SENB only). To ensure plane strain constraint the thickness, B and crack length, a, must satisfy

$$B > 2.5\left(\frac{K_{IC}}{\sigma_0}\right)^2$$

$$a > 2.5\left(\frac{K_{IC}}{\sigma_0}\right)^2,$$

where σ_0 is the yield stress of the material. As will be discussed later (see Sect. 7.6.4) these requirements insure that B and a are large relative to the expected crack tip plastic zone size.

Fracture toughness makes sense only if the toughness is measured from a sharp crack. To start a sharp crack, the test specimen is machined with either a V-notch as shown in the specimens of Fig. 6.1, or with a chevron-notch as discussed in Sect. 6.5. The test specimen is loaded cyclically to initiate and grow a fatigue crack from the notch. The fatigue crack should be grown for approximately 1.3 mm with the last half of the crack growth occurring at a maximum stress intensity factor that is less than approximately 60% of the anticipated K_{IC} value. This insures that the fatigue crack has grown with minimal plasticity and blunting, thus producing a sharp crack. The final total crack length should be $a = (0.45–0.55)W$. The length of the fatigue crack can be measured after the test by examining the fracture surfaces. The fatigued region will show a relatively smooth, flat fracture surface.

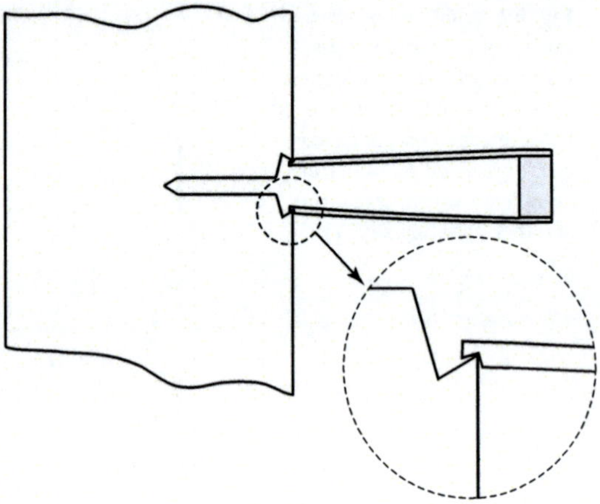

Fig. 6.2 Clip-in gauge for the measurement of the CMOD, v_C. Such gauges can be purchased commercially or built following the detailed description given in E-399. The working principle of the gauge is to measure the bending strain in the gauge arms using strain gauges bonded to the top and bottom surfaces of the gauge. These 4 gauges can be connected into a full Wheatstone bridge

6.2.2 Equipment

To perform the test a calibrated tension/compression testing machine with capability for recording load and displacement signals must be needed. Needless to say, careful calibration of the testing machine and load cells is critical to the accurate determination of K_{IC}. To sense the initiation of fracture the crack mouth opening displacement (CMOD) (labeled as v_C in Table 5.1) must be measured. This is accomplished with a clip-in gauge that mounts to knife edges either machined into or screwed onto the test specimen as shown in Fig. 6.2. Designs for the loading fixtures are given in the standard. For the SENB sample the fixture consists of two rollers for supporting the sample from below and a roller used to load the sample from above.

6.2.3 Test Procedure and Data Reduction

Once the test sample has been prefatigued and specimen dimensions carefully measured and recorded, the sample is loaded steadily at a rate such that K_I increases at 0.55 to 2.75 MPa$\sqrt{\text{m}}$/s. During the test record the load, P and CMOD, v_c. Continue the test until the maximum load is reached.

Typical load-CMOD plots are shown in Fig. 6.3. To perform the data reduction a line is first drawn along the initial part of the P–v_C curve. Then a line from (0,0) is drawn with a slope of 0.95 of the first line. This is the offset line. The following procedures are used to determine a valid K_{IC}. Following the rules given in the caption to Fig. 6.3 determine P_Q. If $P_Q/P_{max} < 1.1$ then calculate a provisional toughness value, $K_Q = P_Q f(a/W,\ldots)$, where the equations for $f(a/W,\ldots)$ are given in Table 5.1. Then check the constraint conditions, i.e. if $2.5(K_Q/\sigma_0)^2 < \min(B,a)$, then $K_{IC} = K_Q$. If these conditions are not satisfied increase B by 50% and re-run the test.

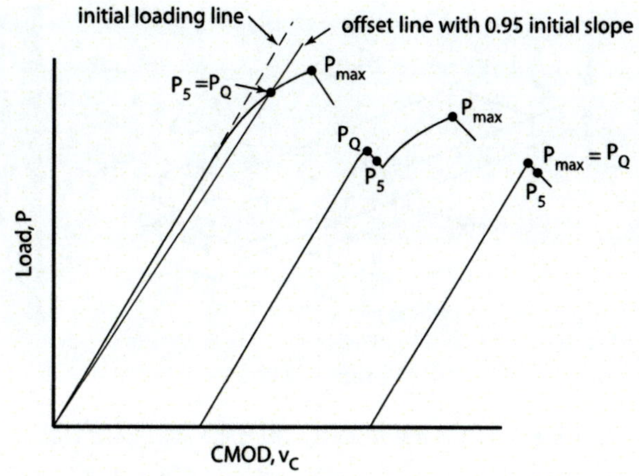

Fig. 6.3 Load-displacement records. The *dashed line* has a slope equal to the P–v slope during initial part of loading. Offset line (*solid*) has slope of 0.95 of the slope of the P–v curve during initial part of loading. P_5 is the load value where the 0.95 offset line intersects the P–v data. $P_Q = \{P_5$ or max P reached before $P_5\}$. P_Q is the load used to establish K_Q, the provisional toughness value

6.3 Interlaminar Fracture Toughness Tests[1]

Composite materials can be prone to interlaminar fracture, i.e. failure between the plys, or layers, that make up a laminated structure. Such fractures may arise from impact or from stresses on pre-existing flaws due to shear loading (leading to Mode-II fracture) or buckling of the layers under compressional loading (leading to Mode-I fracture). Reliable data for the Mode-I and Mode-II interlaminar toughness is thus an important characterization of a composite laminate. The ASTM standard Mode-I test for interlaminar toughness will be discussed in detail. Procedures for Mode-II and mixed Mode testing are reviewed briefly.

6.3.1 The Double Cantilever Beam Test

The DCB test is the most widely used test for determination of Mode-I interlaminar fracture toughness. The DCB test is standardized by ASTM for testing of unidirectional composites [3]; however, it has also been extensively used for testing of woven fabric composites [4, 5].

6.3.1.1 Geometry and Test Procedure

The geometry of the DCB test is shown in Fig. 6.4. The test uses a rectangular specimen, with constant thickness h, width B, and length L. The specimen contains a pre-implanted, non-adhesive insert which serves as an initial delamination. The opening loading (mode I) is applied with loading blocks bonded at the cracked end of the specimen. The test is performed in a testing machine for which the applied load and displacement data can be recorded in real-time.

[1]This section contributed by Michael Czabaj.

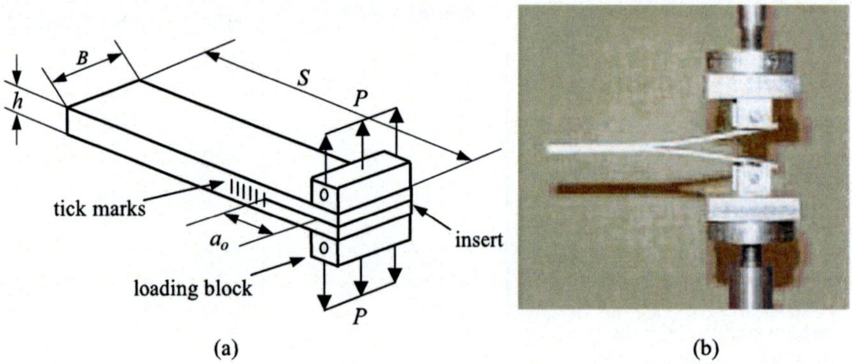

Fig. 6.4 (**a**) DCB test geometry. (**b**) Photo of a typical DCB test underway

The loading is applied in the displacement control, with rates typically between 0.02 to 0.2 in/min. Upon fracture the specimen is unloaded to roughly 90% of the critical load, and the crack growth increment is measured. This measurement is aided with equally spaced tick marks painted on both sides of the specimen, ahead of the initial crack tip. If the difference between measurements from both sides exceeds 0.1 in, the test must be stopped and specimen realigned in the fixture. Since the DCB test is inherently stable the growth increment is small. The specimen can then be fully unloaded and the above procedure repeated for several more crack growth increments. Once the test is over, the specimen is examined for any permanent deformation and ideally split open to examine the fracture surface. For some epoxy matrix composites, propagation of a crack leaves a noticeable imprint at each successive load-up, and can be used to correct the crack extension values measured during the test.

When manufacturing the test specimen is it suggested that the delamination insert thickness be no greater then 0.0005 in, providing a fairly sharp crack tip. The specimen should be spray painted with a silver paint on both sides, and tick marks should be marked with pencil in increments of 1/32 in ahead of the crack tip. The loading tabs must be adhered to the specimen with a glue that will not debond under the maximum applied load. Special care must be taken to align the loading tabs with the specimen, since any misalignments may result in a non-uniform crack growth.

6.3.1.2 Data Reduction Methods

There are number of data reduction methods available for the DCB test; however, the two most common methods are the modified beam theory (MBT) method and compliance calibration (CC). The MBT approach assumes that classical beam theory can be used to determine stresses and deformations, and that both arms of the DCB specimen can be modeled as cantilever beams. Using this, one can derive an expression for a tip deflection, v of one DCB specimen arm given by

$$v = \frac{Pa^3}{3EI}, \tag{6.1}$$

where E is the flexural modulus of elasticity, and I is the area moment of inertia of a single arm. Let $v_L = 2v$, where V_L is the total displacement between the loading pins. The compliance of the entire specimen is therefore

$$C = \frac{v_L}{P} = \frac{2v}{P} = \frac{2a^3}{3EI}. \tag{6.2}$$

Differentiating (6.2) with respect to crack length, and substituting into Eq. (3.25) yields

$$G = \frac{P^2a^2}{BEI} \tag{6.3}$$

which can then be made independent of E and I, by substitution of Eq. (6.1) into Eq. (6.3). The final result yields an expression for the mode I fracture toughness

$$G_{Ic} = \frac{3P_cV_{Lc}}{2Ba} \tag{6.4}$$

which depends only on the specimen geometry, critical load, P_c, and displacement δ_c. As stated in [3], Eq. (6.4) will overestimate G_{Ic} because in practice the arms of the DCB specimen will undergo a finite rotation at the delamination front [3, 6]. To alleviate this problem, the crack length used in Eq. (6.4) is increased by a correction factor, Δ, giving a_{true} as a $+\Delta$. This new crack length is then substituted into Eq. (6.2) along with an expression for $I = Bh^3/12$, giving

$$C = \frac{8(a+\Delta)^3}{EBh^3}. \tag{6.5}$$

Now, by plotting the cube root of the experimentally determined compliance at different crack lengths (Fig. 6.5), one can see that if no correction is required then the plot will go through the origin. If the plot has an x-axis intercept, then the magnitude of that intercept is used as a correction factor Δ. In certain instances the direct approach of the MTB may ignore the effects of shear deformation, which is manifested by a non-linear relationship between compliance and crack length when plotted on the log-log scale. In this case, a better measure of fracture toughness can be obtained with the compliance calibration method [3, 7]. With this method, the compliance versus respective crack length can be plotted and fit with the following expression

$$C = Ra^n. \tag{6.6}$$

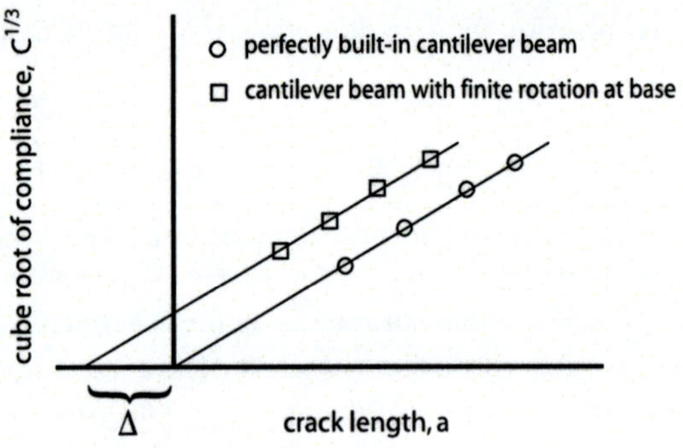

Fig. 6.5 Effect of finite rotation of the DCB specimen

This expression is then substituted into Eq. (3.25), resulting in

$$G = \frac{P^2}{2B} nRa^{n-1}. \tag{6.7}$$

Next, solving Eq. (6.6) for R, substituting into Eq. (6.7) and noting that $C = P/\delta$, the CC expression for mode I fracture is

$$G_{Ic} = \frac{nP_c v_{Lc}}{2Ba}. \tag{6.8}$$

Equations (6.4) and (6.8) are accurate for specimens with low mode I toughness; however, they must be further corrected for toughened composites which may undergo significant deformation prior to fracture [3, 6, 8, 9]. It is suggested that these corrections are made if the ratio of $V_{Lc}/2$ to the crack length, a, exceeds a value of approximately 0.4. Large deformation results in a shortening of the moment arms and a resisting moment caused by the rotation of the loading tabs.

6.3.1.3 Example Results

An example of typical experimental data for a DCB test is presented in Fig. 6.6. The material tested was an IM7-8551, unidirectional, 24-ply graphite/epoxy composite. The specimen was cut to a width of 1 in, and the average thickness of the uncracked region was 0.118 in. A 0.0005 in, Teflon film was used as a starter crack. The loading blocks were machined out of aluminum, with a total width of 0.5 in, and a height of 0.375 in. Since each test appears to be linear to fracture in Fig. 6.6, there is only one definition of a critical load and displacement. The large displacement corrections were unnecessary, since the largest $\delta/2a \ll 0.4$. Figure 6.7(a) depicts the compliance plotted versus the crack length as fit with Eq. (6.6).

The resistance curve (R-curve), or fracture toughness plotted with crack length is presented for all data points in Fig. 6.7(b). There is a good correlation between the MBT and CC methods. In Fig. 6.7(b), the toughness value obtained from the initial crack length is the highest, and it decreases with increasing crack length. The first value, referred to as the non-precracked (NP) value, initiates from the delamination

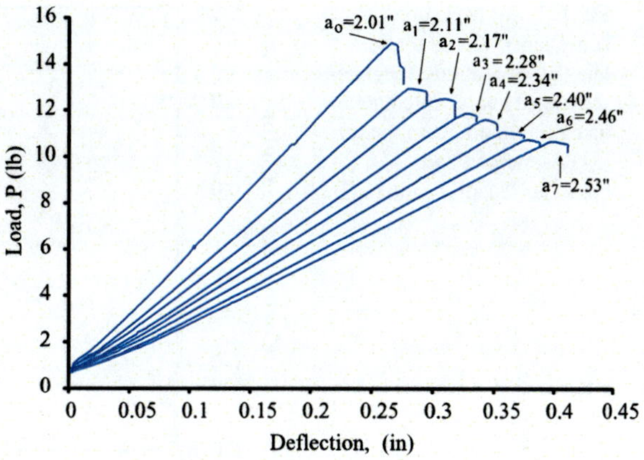

Fig. 6.6 Load-deflection data from test of 24-ply graphite/epoxy composite

insert that makes a relatively blunt crack tip (typically associated with a pocket of resin just ahead of the insert). Toughness on the second load-up, referred to as the precracked (PC) value, dramatically decreases in magnitude due to initiation from a sharper, more naturally occurring crack tip.

6.3.2 The End Notch Flexure Test

The end notch flexure test (ENF) is the most common Mode-II interlaminar fracture toughness test, likely to be standardized by ASTM in near future. The geometry of the ENF test is shown in Fig. 6.8(a). The specimen is placed on top of two supporting rollers with span of 2L. The bending load P is applied via a loading roller located in the mid-span. The crack tip is aligned such that it falls between the loading and supporting rollers, in the area of constant internal shear force. The shear force results in longitudinal sliding of the crack faces, resulting in pure mode II loading.

Because of the unstable nature of the test, upon fracture the crack tip travels and arrests underneath the loading roller. If one wishes to obtain an additional toughness value from the same specimen, the crack front must be non-destructively examined, and straightened if necessary [10]. Prior to the fracture test, it is recommended to obtain a specimen's compliance as a function of crack length, such that it can be used in the compliance calibration method of data reduction.

There have been over a dozen proposed data reduction approaches for the ENF test, summarized in [11]. From FE analysis [11], it appears that the most accurate measure of G_{IIc} is obtained with the compliance calibration method. The compliance data is best fit to

$$C = A + ma^3. \tag{6.9}$$

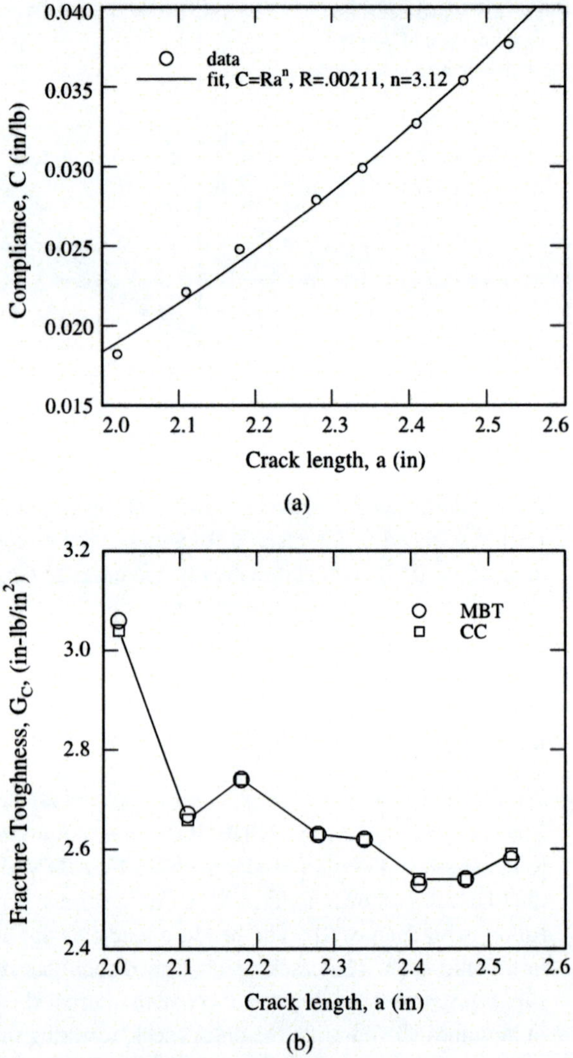

Fig. 6.7 (**a**) Measured compliance versus crack length. (**b**) Fracture toughness resistance curve, calculated using modified beam theory (MTB) method and compliance calibration (CC)

Substituting Eq. (6.9) into Eq. (3.25) gives the following expression for mode II fracture

$$G_{IIc} = \frac{3mP_c^2a^2}{2B}. \tag{6.10}$$

6.3.3 Single Leg Bending Test

There have been a number of mixed Mode I-II tests proposed, with the Mixed Mode Bending (MMB) test and the Single Leg Bending (SLB) test being the most popular.

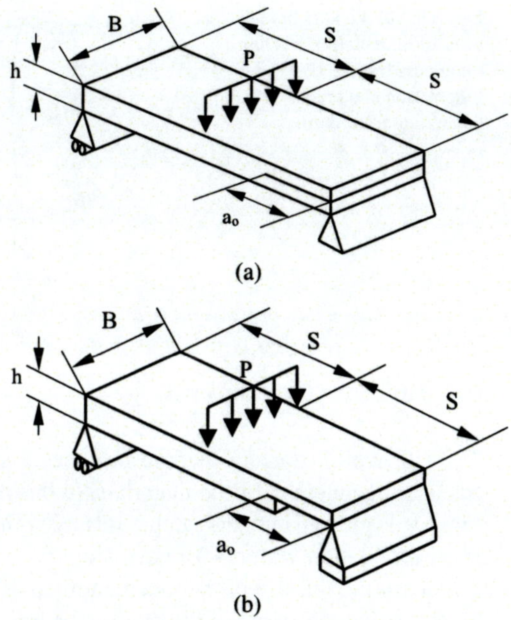

Fig. 6.8 (**a**) End notch flexure (ENF) test geometry for Mode-II interlaminar toughness testing. (**b**) Single leg bending (SLB) test geometry for mixed-Mode interlaminar toughness testing

The MMB test is standardized by ASTM [12], and is popular due to the fact that a single specimen can be tested under a full range of mode-mixity (the mode-mixity is defined as the fraction of mode II ERR, to the total ERR). The MMB test, however, is fairly difficult to run, and has a complicated data reduction scheme influenced by number of geometrical non-linearities [13]. On the other hand, the SLB test provides only a single mode mixity of 0.4, but the test procedure and data reduction is very similar to that of the ENF test. In recent work, [14] outlines a method to accurately interpolate any mode I-II mixity from the DCB, SLB and ENF data, negating the advantage of the MMB test.

Figure 6.8(b) depicts a typical SLB test. The specimen geometry is the same as for the ENF test; however, the SLB specimen has the lower of the two legs removed, causing a transfer of the total load to the upper leg. Similar to the ENF test, a compliance calibration can be performed and used with Eqs. (3.25) and (6.9) to obtained the critical fracture toughness. One can also reduce the data using the classical plate theory formulation [13], given by

$$G_{c_{SLB}} = \frac{3P_c\delta_c a^2}{2B}\left[\frac{D/D_T - 1}{2S^3 + a^3(D/D_T - 1)}\right], \tag{6.11}$$

where D is the uncracked bending stiffness, and D_T is the bending stiffness of the top leg.

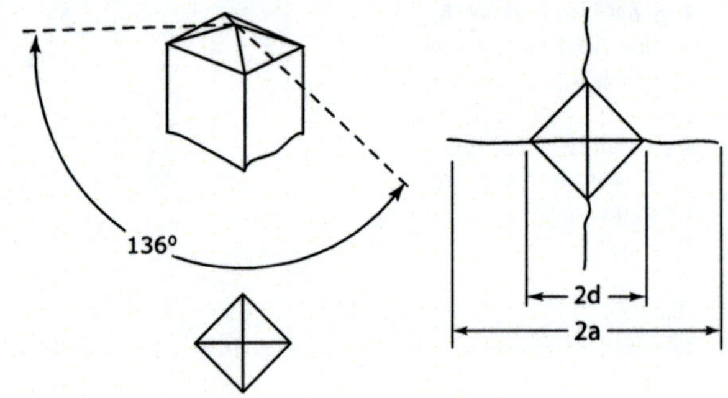

Fig. 6.9 (**a**) Vickers indenter used for indentation fracture toughness testing. (**b**) Indentation and cracks emanating from corners

6.4 Indentation Method

The indentation fracture method provides a simple and fast measurement of the fracture toughness of brittle materials. In this method a sharp, Vickers indenter, see Fig. 6.9, is pressed onto the sample surface with a weight of 200–3000 g. This results in an indentation with cracks emanating from the corners, as shown in the figure. By measuring the size of the indentation, to determine the material's hardness, and by measuring the length of the cracks, the fracture toughness can be estimated.

A detailed description of the mechanics and experimental observations related to indentation fracture can be found in Lawn et al. [15]. The cracks formed during indentation can be considered as a pair of orthogonal half-penny cracks, diameter $2a$, wedged open at their center by a point force. The point force is due to the residual stress field left behind due to inelastic deformation at the tip of the indenter. From dimensional analysis the stress intensity factor must be

$$K_I = \frac{\chi P}{a^{3/2}}. \tag{6.12}$$

The constant χ is a function of the indenter geometry, and the material's hardness. From Eq. (18) of Lawn et al.

$$\chi = \xi_0 (\cot \psi)^{2/3} \left(\frac{E}{H_T} \right)^{1/2}, \tag{6.13}$$

where the constant $\xi_0 = 0.032 \pm 0.002$ is found by fitting experimental data, ψ is the half angle of the indenter, 68°, E is the Young's modulus, and H_T is the hardness. The Vicker's hardness is found from

$$H_v = 1.854 \frac{P/g}{(2d)^2}, \tag{6.14}$$

where P/g is the indentation mass (kg), g is the acceleration of gravity and $2d$ is the indentation diagonal (mm). This is converted to true hardness in GPa by [16]

$$H_T = 1.057 \times 10^{-2} H_v. \tag{6.15}$$

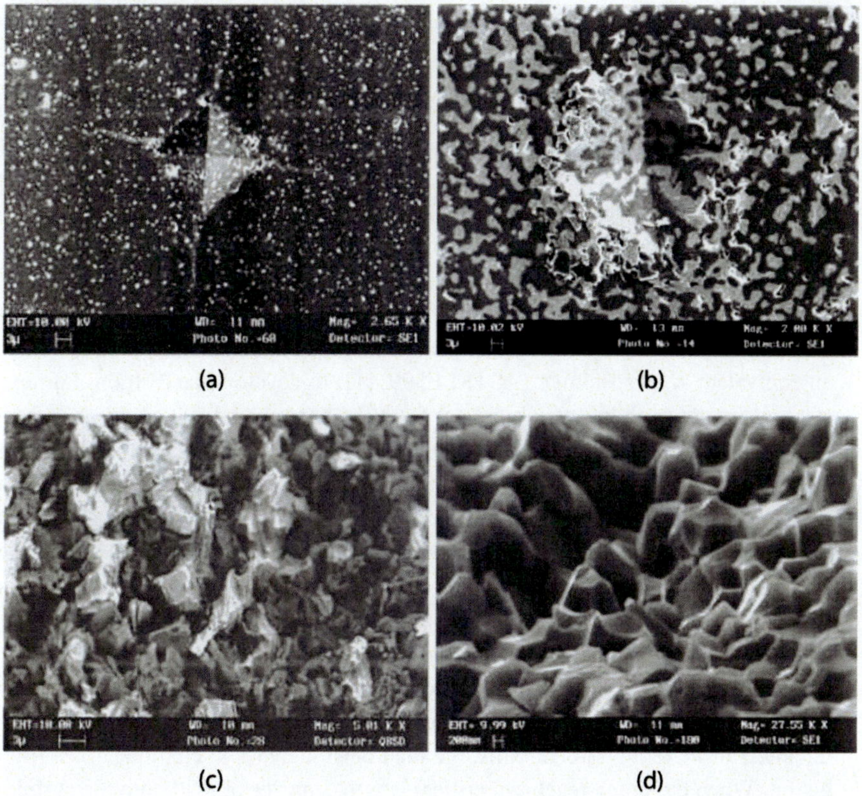

Fig. 6.10 (**a**) Indentation and indentation crack formed using Vickers indenter on a metal-ceramic composite made by a sol-gel method with 95% V_f Al_2O_3, 5% V_f Ni. (**b**) Attempt to measure fracture toughness in a more ductile material failed. Shown is an indentation formed in a composite made by a sol-gel method with 67% V_f Al_2O_3, 33% V_f Ni. (**c**) Fracture surface of material with 90% V_f Al_2O_3, 10% V_f Ni. The surface shows intergranular fracture of the alumina as well as some ductile deformation of the Ni. (**d**) Fracture surface of material with 50% V_f Al_2O_3, 50% V_f Ni. Surface shows extensive ductile deformation of the Ni phase

In Eq. (6.12), P is the applied force in N.

An example of an indentation crack in a metal-ceramic composite is shown in Fig. 6.10a. This material was made by a sol-gel method [17] and is 95% by volume ceramic (Al_2O_3) and 5% by volume metal (Ni). A clearly defined indentation with cracks emanating from all four corners can be seen in the figure. This material has a fracture toughness on the order of 3.5 MPa$\sqrt{\text{m}}$. However, when the volume fraction of the metal is increased to 33%, Fig. 6.10b, cracks are not formed at the indenter corners, the material is simply indented, providing a measure of the hardness but not of the fracture toughness. In such cases, other methods such as the chevron-notch method discussed below must be used.

6.5 Chevron-Notch Method

Chevron notched (CN) test specimens [18–23] are used extensively for measuring fracture toughness of ceramics and other brittle materials for which the standard toughness measurement requiring precracking is difficult to implement. Different specimen shapes such as four- and three-point bend samples, Fig. 6.11, can be used for non-standard tests while the ASTM E 1304 standard test [23] use rod, shown in Fig. 6.12, or bar specimens. In each case a V-shaped notch is cut into the sample to act as a crack started upon loading. A CN test specimen has several advantages over other fracture toughness specimens. First, it needs to be only half the size of an equivalent K_{IC} specimen (ASTM E399, [1]) to develop plane strain. Further, it requires no precracking, which frees it from the need for fatigue equipment, which can be expensive. In addition, the method is particularly attractive for brittle materials, because crack in a CN specimen initiate generally initiate stably without specially designed fixtures. The entire crack growth is generally stable, allowing the load-displacement curve to be recorded and the work of fracture to be calculated. The chevron-notch method and variations are discussed here. As with the method for determination of K_{IC}, if you wish to determine valid fracture toughness values you will need to read and carefully follow the standard E 1304 method [23].

In CN samples, the crack initiates from the apex of the notch at a low load due to the concentration of stress there. As the crack advances, it grows into an increasing thickness of material, thus initially, the load must increase to continue crack propagation. When the crack reaches a critical length, transfer of load onto the shrinking remaining ligament requires a decreasing load for stable crack growth. In chevron-notched samples, for a given load, the stress intensity factor is theoretically infinite for $a = a_0$, where a_0 is the initial crack length, then decreases with increasing crack length, and finally increases again.

For example the normalized stress intensity factor, $Y \equiv K_I B\sqrt{W}/P$ (where K_I is the mode-I stress intensity factor, B is the specimen width, W is the height, and P is the load) for a three point bend specimen, calculated using Wu's solution [20], is plotted versus normalized crack length, $\alpha \equiv a/W$, in Fig. 6.13 for a number of different notch depths, α_0. In each case Y is initially infinite, drops with increasing crack growth until the critical crack length, $\alpha = \alpha_c$ is reached, and then increases. Note that α_c is a function of α_0.

Define K_{IC} to be the fracture initiation toughness, and $K_R(\Delta a)$ to be the fracture propagation toughness as a function of crack growth, or "R-curve", with $K_R(0) = K_{IC}$, and $\Delta a \equiv a - a_0$.

Assuming for now that $K_R = \text{constant} = K_{IC}$, as the crack grows, the load must initially increase to maintain $K_I = K_{IC}$, and then drop after the crack reaches its critical length, resulting in a load-displacement record similar to the example in Fig. 6.15 obtained for a small ceramic sample tested in three-point bending with the setup shown.

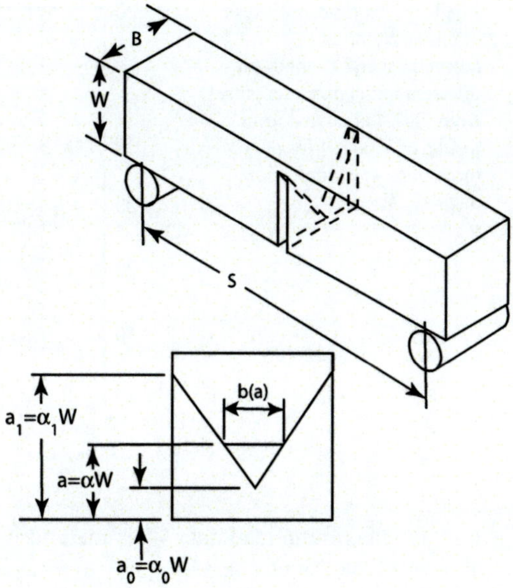

Fig. 6.11 Chevron-notched bend test specimen, from [24]

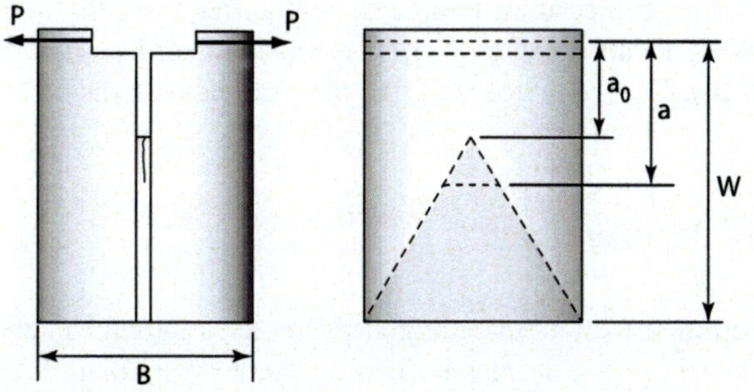

Fig. 6.12 Rod specimen for ASTM E-1304, "Standard test method for plane strain (chevron-notch) fracture toughness of metallic materials." Specimen size restrictions: $B > 1.25(K_{IV}/\sigma_0)^2$. $W/B = 1.45$ or $W/B = 2.0$. $a_0 = 0.481B$ (short rod) or $a_0 = 0.400B$. Load and displacement at the load line are measured are recorded in the standard test. A rectangular specimen can also be used. Further details may be found in [23]

6.5.1 K_{IVM} Measurement

Data such as shown in Fig. 6.15b are used in three ways to calculate fracture toughness. In the first method, which is part of the E-1304 standard, the minimum value of Y, $Y_c \equiv Y(\alpha_c)$, is calculated for the particular test specimen geometry. The fracture toughness is then calculated by

$$K_{IVM} = P_{max} Y_c / B\sqrt{W}, \tag{6.16}$$

where P_{max} is the maximum load reached in the fracture test. The notation K_{IVM} is used to indicate that the toughness was derived from a chevron notched test using

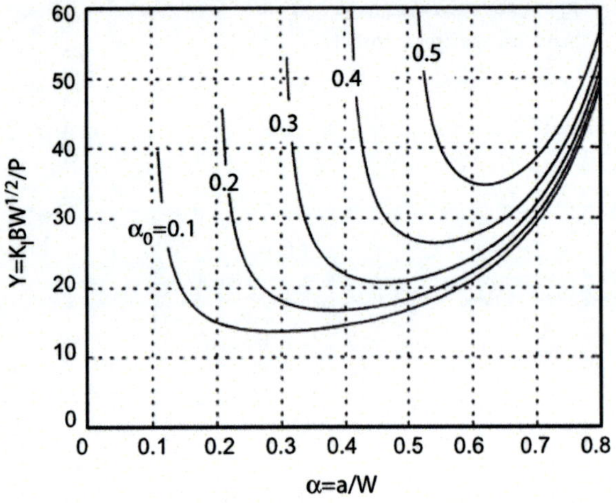

Fig. 6.13 Normalized stress intensity factor vs. normalized crack length for different initial crack lengths, from [24]. Calculated from results of Wu [20] for a three-point specimen with $S = 4W$, $W = B$

only the maximum load data. This method is very simple to use if a calculation of Y_c is available, since one needs only to measure the test specimen geometry and the peak load to calculate K_{IVM}. Note that since the crack must grow a finite distance before $\alpha = \alpha_c$, the measured toughness will differ from the initiation toughness, K_{IC}, unless K_I is independent of crack extension, i.e., $K_{IVM} = K_{IC}$ only if $K_R = K_{IC}$ for all Δa. Generally you will find that for real materials $K_{IVM} \neq K_{IC}$.

6.5.2 K_{IV} Measurement

In the second method two unloading steps are used in order to ensure that the material behavior is primarily elastic. During the test the load, P and load line displacement, v_L, (i.e. crack opening displacement measured at points that are in line with the load application) are measured and recorded. The resulting record will look something like the graph in Fig. 6.14.

Let's think of running two tests. In the first the test is run to complete fracture with no unloading. This test will be used to establish unloading points for the next test. During the initial part of the loading the curve is linear with slope (stiffness) m_0. Based on analysis of the test specimen, the ratio of the initial stiffness to the stiffness that the specimen would have if $a = a_c$ (crack length where Y is minimum) can be calculated. In terms of compliance this slope ratio is $r_c = C(a_0)/C(a_c) < 1$. Draw a line on the $P - v_L$ plot with slope $m_0 r_c$. Now draw two lines with slopes $0.8 m_0 r_c$ and $1.2 m_0 r_c$. Determine the loads, P_1 and P_2 where these two lines intersect the P–v_L curve. In the second test load the specimen to P_1 and then unload it to 3% to 10% of P_1. Reload and continue the test until the peak force is reached and the load drops back to P_2. Unload and reload as before and continue the test. The resulting record will look something like the graph in Fig. 6.14.

To validate the data a number of quantities must be computed. First determine the average force between the two unloading lines. Draw a horizontal line at this force

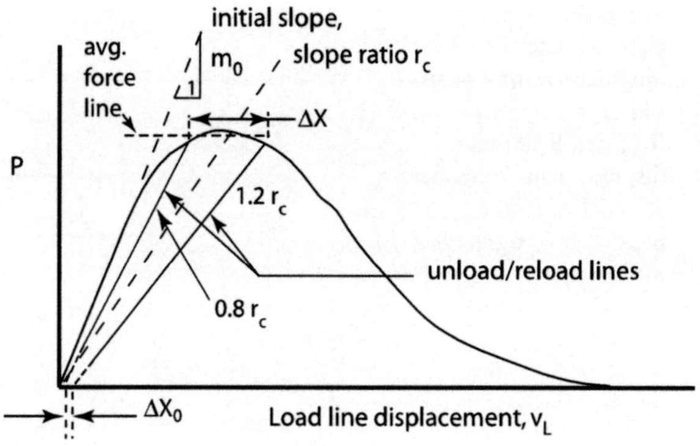

Fig. 6.14 Load-displacement record for chevron notch test

level. Determine the points where the extensions of the two unloading lines intersect the average force line. The horizontal distance between these intersections is ΔX. Extend the unloading lines so that they intersect $P = 0$. The horizontal distance between these intersections is ΔX_0. Determine the maximum force, P_{max} reached in the test. Determine the force P_C where the line with slope ratio r_C intersects the P–v_L curve. Calculate a provisional toughness value,

$$K_{QV} = \frac{Y^* P_C}{B\sqrt{W}}.$$

As with the E-399 test a number of criteria must be satisfied before the provisional value can be considered a valid toughness. IF $P_{max} < 1.1 P_C$ and $-0.05 < \Delta X_0/\Delta X < 0.10$ and $B \geq (K_{QV}/\sigma_0)^2$, THEN

$$K_{QV} = K_{IV}.$$

6.5.3 Work of Fracture Approach

In the third method, the load-displacement curve is integrated to find the "work of fracture", U_f,

$$U_f \equiv \int_0^{v_{Lf}} P dv_L, \tag{6.17}$$

where v_L is the load point displacement and v_{Lf} is the displacement at complete fracture. Equating the total work done to the fracture surface energy, the fracture toughness G_C (the area averaged energy release rate) is

$$G_C = \frac{U_f}{A_f}, \tag{6.18}$$

where A_f is the area of one fracture surface.

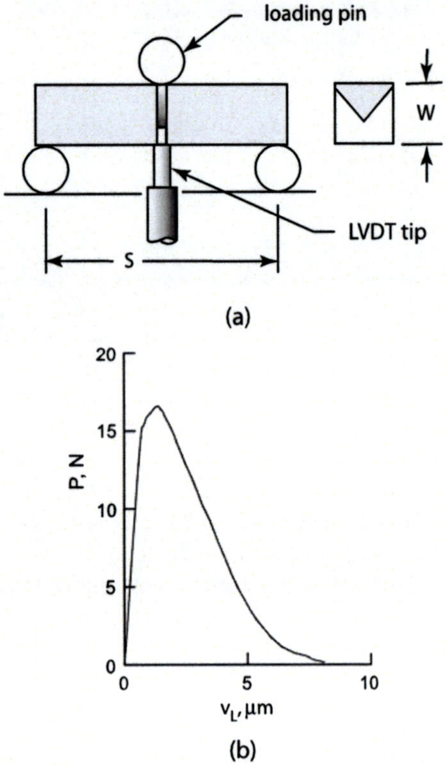

Fig. 6.15 (**a**) Setup for chevron-notch fracture toughness testing of small samples of alumina (Al_2O_3). (**b**) Load-load point displacement. Toughness K_{IVM} is measured based on peak load. Toughness K_{wof} is measured based on $\int P dv_L$

Assuming plane-strain, small-scale yielding conditions, the relationship between the stress intensity factor, K_I, and the energy release rate, G is

$$G = \frac{K_I^2(1-\nu^2)}{E}, \tag{6.19}$$

where ν is the Poisson's ratio, and E is the Young's modulus. Thus, a "work of fracture" value of toughness can be calculated as

$$K_{wof} = \sqrt{\frac{G_C E}{1-\nu^2}} = \sqrt{\frac{EU_f}{(1-\nu^2)A_f}}. \tag{6.20}$$

Note that $K_{wof} = K_{IVM} = K_{IV} = K_{IC}$ only in the case of materials with a flat R-curve. When the material is somewhat ductile it may have a rising R-curve and K_{IVM} and K_{wof} will diverge. Such is the case for nickel-alumina composites of varying metal volume fraction [24]. As shown in Fig. 6.16a the toughness as measured by K_{IV} increases only a small amount with increasing metal volume fraction, while the toughness measured as K_{wof} increases by a factor of two as the metal volume fraction increases from 5 to 50%. The fracture surfaces of these metal-ceramic composites, Figs. 6.10c, d show that at low metal fraction the fail-

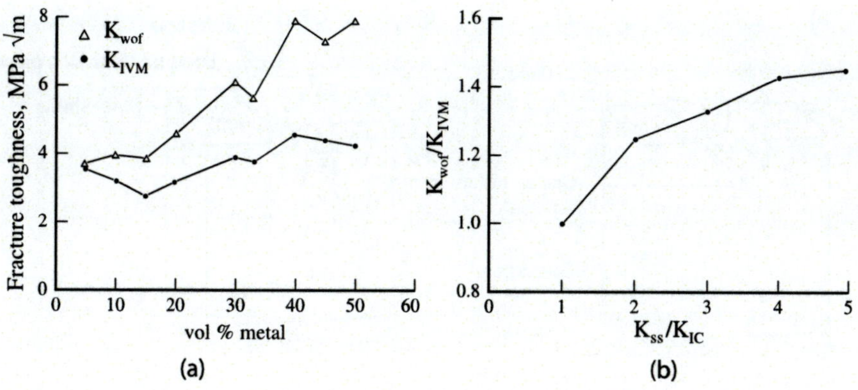

Fig. 6.16 (**a**) Fracture toughness of nickel-alumina composites of varying metal volume fraction. (**b**) K_{wof}/K_{IVM} vs. K_{ss}/K_{IC} calculated for a material with an *R*-curve that rises from K_{IC} to K_{ss} over a distance of 1 mm in a chevron-notched three-point bend sample with $W = B = 2mm$. From [24]

ure is brittle with cracks that grow around the ceramic grains. At higher volume fraction the cracks still grow around the grains, but the metal phase has deformed significantly and in so doing contributed to the energy required to fracture the sample, increasing its K_{wof}. A simple analysis of these experiments, whose results are shown in Fig. 6.16b, shows the extent to which K_{IVM} and K_{wof} diverge as the steady state value of the crack growth resistance curve increases.

An additional concern with chevron-notched samples is that the crack growth is not always stable. Due to the finite notch thickness the initial stress intensity factor is not infinite and thus a finite load is needed to start the crack. Depending on the relative stiffness of the loading system relative to the specimen the crack may grow stably, may "pop-in", i.e. grow unstably for a very short length and then stabilize, or may grow unstably across the entire sample. The ASTM standard [23] gives procedures for the reduction of data in the case that the crack grows in short unstable jumps. However, in the case that unstable growth across the sample occurs insufficient data will be obtained for the determination of either K_{IVM} or K_{wof} and the system must be modified to provide more stable loading.

6.6 Wedge Splitting Method

The wedge splitting method [25–27] is one means of stabilizing crack growth. In addition, wedge loading produces compressive stresses parallel to the crack, stabilizing crack growth direction. The method can be adapted to study a great many materials, including ceramics, rocks, asphalt, concrete, wood and bonds between various materials, can be adapted to study fracture in the presence of bi-axial stress states and can be used at high and low temperatures.

The principle of the method is shown in Fig. 6.17. In a compression testing machine samples with a rectangular groove and a starter notch at the bottom of the

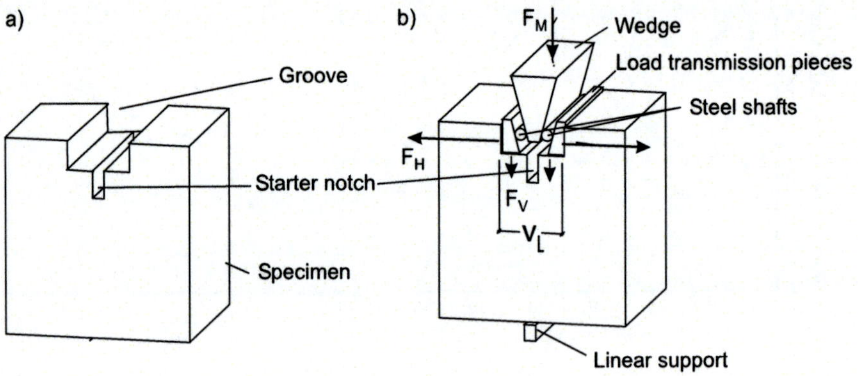

Fig. 6.17 Principle of wedge splitting test method, courtesy of Prof. E.K. Tschegg

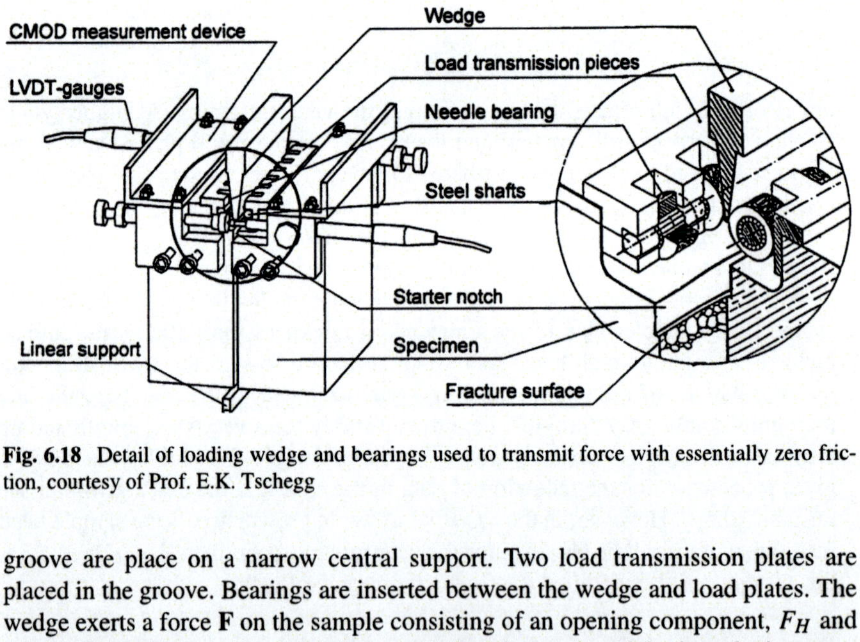

Fig. 6.18 Detail of loading wedge and bearings used to transmit force with essentially zero friction, courtesy of Prof. E.K. Tschegg

groove are place on a narrow central support. Two load transmission plates are placed in the groove. Bearings are inserted between the wedge and load plates. The wedge exerts a force **F** on the sample consisting of an opening component, F_H and a vertical component, F_V, that helps to stabilize the crack path by introducing compression into the sample. Using a small wedge angle tremendous leverage can be obtained. In addition, the loading is very stiff and stable crack growth is achieved in most cases.

As the wedge is driven down, the crack will grow from the starter notch and F_V and the load-point opening displacement are measured and recorded. Figure 6.18 shows the details of the wedge and bearing system used to load the sample. Carefully calibrated LVDTs, positioned directly in line with the point of load application, are used to measure the opening displacement. Using two LVDTs and averaging the result will minimize errors associated with any asymmetry in the loading or geometry. The opening force, F_H is easily calculated from the wedge angle as long

Fig. 6.19 Load-displacement record from a wedge splitting test of a semi-circular limestone sample. Data from Prof. E.K. Tschegg

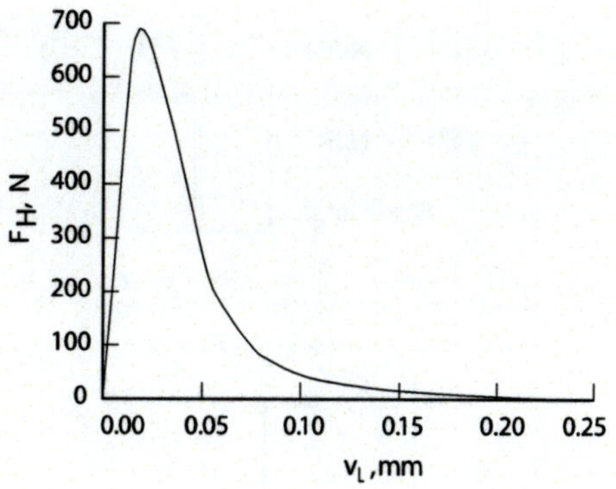

a bearing are used to essentially eliminate friction. Plotting F_H vs. v_L thus displays the opening force and displacement; such plots are similar to those of the chevron-notch method, e.g. Fig. 6.15.

As with the chevron-notch method the load-displacement may be integrated, and the fracture toughness G_C can be calculated as

$$G_C = \frac{\int_0^{v_{Lf}} F_H dv_L}{A_f}, \tag{6.21}$$

where v_L is the opening displacement measured at the point of application of the wedging forces, v_{Lf} is the opening displacement at final fracture and A_f is the fracture surface area. If desired, this can be converted to a K_{wof} using Eq. (6.19). For example, the F_H vs v_L data shown in Fig. 6.19 were obtained from a wedge splitting test of limestone ($E \approx 66$ GPa), using the semi-circular geometry shown in Fig. 3.10 with radius 50 mm, thickness 20.7 mm, initial crack length of 14 mm and crack propagation distance of $(50 - 14) = 36$ mm. Integrating the curve the total fracture energy is 34 N-mm, corresponding to fracture energy of $G_C = 34/(36 \cdot 20.7) = 0.039$ N/mm $= 39$ J/m^2. The corresponding K_{wof} value is $K_{wof} = \sqrt{G_C E/(1-\nu^2)} = 1.66$ MPa$\sqrt{\text{m}}$ (taking $\nu = 0.25$). Note that the value of E is determined from the load-displacement record based on a finite element analysis of the stiffness prior to crack growth. One can compute the stiffness with a reference value of E, say $E = 100$ GPa, and then compare the measured and computed stiffnesses to determine the actual E.

The wedge-splitting method can be used with samples of various types and shapes, as shown in Fig. 6.20. Particularly interesting are samples consisting of two materials bonded together along the prospective crack line. For example the toughness of the bond between two types of asphalt was measured as a function of temperature [28]. Additional compressive stresses parallel to the crack can be applied to measure the fracture toughness under biaxial loading conditions [29].

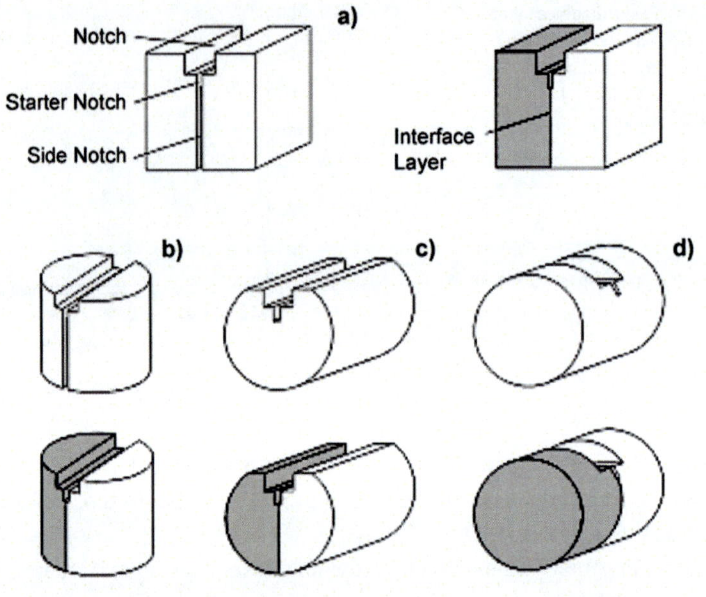

Fig. 6.20 Types of samples used in wedge splitting tests, courtesy of Prof. E.K. Tschegg

6.7 *K–R* Curve Determination

As discussed in Sect. 4.3 the stability of crack growth, i.e. whether a crack will immediately become catastrophic after initiation, or whether it will grow subcritically for some time, allowing corrective action to be taken, depends on the slope of the *R*-curve and on the stiffness of the loading. In the context of LEFM the *R*-curve is characterized in terms of $K_I(\Delta a)$, i.e. the value of Mode-I stress intensity factor needed to sustain crack growth. The measurement of $K-R$ curves is described in ASTM E 561, Standard Practice for R-Curve Determination [30]. Unlike the plane strain and chevron-notch toughness tests, no thickness requirements are given for $K-R$ curve testing, thus the test should be conducted for the thickness of material that will be used in the prospective application.

6.7.1 *Specimens*

Compact tension, center cracked tension and a wedge loaded, edge cracked specimen can be used for *R*-curve testing. For details of the wedge loaded specimen see [30]. Here I will discuss the method in the context of the center crack tension specimen. As a rule of thumb, the specimen width, W should be $W > 14r_p$ where $r_p = \frac{1}{\pi}(\frac{K_{max}}{\sigma_0})^2$ is the estimated plastic zone size (see Sect. 7.6.4 for details), σ_0 is the yield strength of the material and K_{max} is the maximum value of K_I reached in the test. The crack length ($2a$ in Table 5.1) should be $2a = W/3$. A starter notch with a root radius of less than 0.08 mm should be cut to facilitate fatigue pre-cracking. The specimen is to be fatigue pre-cracked according to E-399 [1].

Recall that the T stress for a crack in tension is compressive. In thin specimens this compressive T stress can result in buckling along the crack line. To prevent buckling, "anti-buckling" guides must be constructed and used. The guides consist of thick plates attached to the central portion of the specimen. The plates should constrain buckling but without any restricting in-plane motion due to tensile loading. Teflon sheets, grease, or a heavy oil can be used as lubricants between the specimen and anti-buckling guides.

6.7.2 Equipment

In addition to the testing machine and loading fixtures a means of measuring the crack length is needed.

6.7.2.1 Optical Measurement of Crack Length

In the case that the test specimens are thin enough that the crack front will be nearly perpendicular to the specimen surface optical measurement of crack length can be used. A 30 to 50× microscope attached to a translation stage with micrometers or a digital readout can be used as shown in Fig. 6.21. The microscope is translated until its cross hairs are centered on the crack. Video recording through a low power microscope can be used as well. In both cases resolution of the crack will be enhanced by polishing the surface of the test specimen and by careful attention to lighting. A crack length resolution of at least 0.2 mm is required.

6.7.2.2 Compliance Method for Crack Length

As discussed in Sect. 3.4.2 in an elastic material there is a one-to-one relation between crack length and compliance, hence if compliance can be determined then the crack length can be determined. In the center cracked panel the displacement to be measured is the opening at the center of the crack. This measurement is accomplished through the use of a clip-in gauge inserted into a hole drilled at the center of the specimen as shown in Fig. 6.22. The hole must be beveled to provide knife edges for contact of the displacement gauge with the test specimen.

6.7.2.3 Other Methods for Crack Length

Other methods such as eddy-current probes or electrical resistance measurements, either of the specimen itself or of gauges adhered to the specimen, may be used to measure crack length.

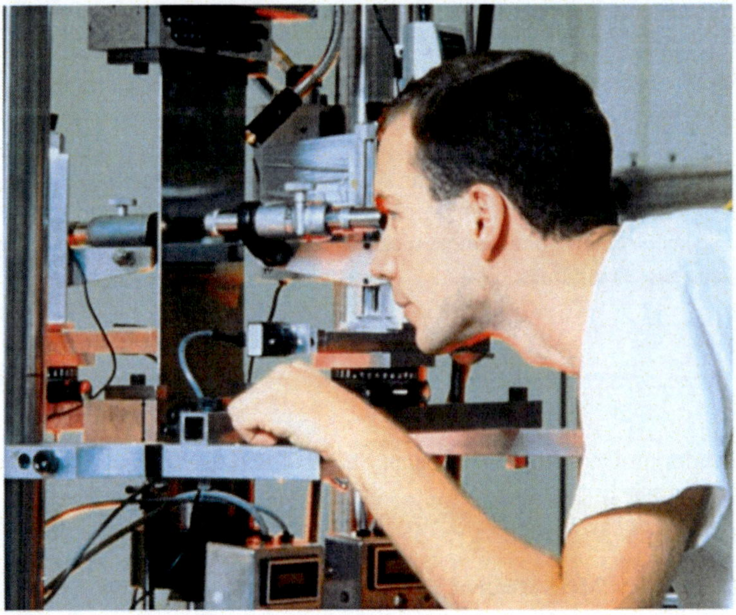

Fig. 6.21 Measurement of crack length using a low power microscope attached to a translation stage. Crack tip position is determined by locating cross hairs on the crack and reading position via a digital encoder attached to the shaft of the translation stage

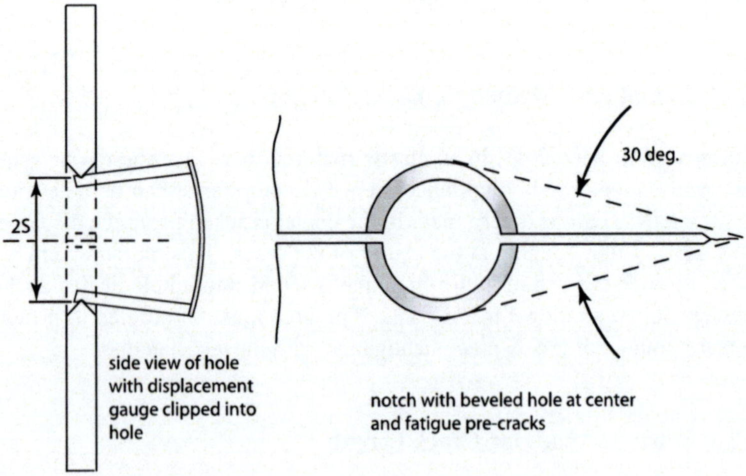

Fig. 6.22 Gauge for measuring crack opening displacement for center cracked panel. Hole size should be small enough so that when $2a \approx W/3$ an envelope of less than 30° is formed from the crack tip to the hole edges

6.7.3 Test Procedure and Data Reduction

6.7.3.1 By Measurement of Load and Crack Length

The specimen should be loaded in steps with crack length measurements made at each step. At least 5 measurements should be made, although 15 or more would provide better results. Crack length should be measured with an accuracy of 0.2 mm or better. For each pair of load and crack length measurements (P, a) calculate the stress intensity factor, K_R using the equation in Table 5.1 with a replaced by a plastic zone adjusted crack length, $a_{eff} = a + r_p/2$ where $r_p = \frac{1}{\pi}(\frac{K_R}{\sigma_0})^2$. As discussed in Sect. 7.2.1 increasing the crack length in this way accounts for the increase in K_I due to crack tip plasticity. Plot K_R vs. $a_{eff} - a_0$ the plasticity corrected crack growth increment.

6.7.3.2 By Measurement of Load and Compliance

If the compliance method is used to determine crack length the procedure is similar. The sample is loaded to the point where crack growth begins. It is then unloaded and reloaded, allowing crack growth to occur between each load/unload cycle. The resulting plot of P vs. v_C should look something like Fig. 6.23. Determine compliance, C for each unloading. Then determine crack length through either an experimental calibration between C and a, or using the analytical expression given in E-561 [30], (this expression is for center cracked panel only)

$$C(a) = \frac{2}{EB}\left[\frac{\pi a}{W}\frac{1}{\sqrt{\sin \pi a/W}}\right]\left[\frac{2W}{\pi S}\cosh^{-1}\left(\frac{\cosh \pi S/W}{\cos \pi a/W}\right) - \frac{1+\nu}{\sqrt{1+\frac{\sin^2 \pi a/W}{\sinh^2 \pi Y/W}}} + \nu\right]\frac{S}{W}, \tag{6.22}$$

where S is the half span of the gauge and the other parameters are defined in Table 5.1 and in the text. Note that this equation gives $C(a)$ and will need to be inverted to calculate a given C. Note also that E can be determined from the initial slope (i.e. prior to crack growth) and known initial crack length. Make all crack length calculations based on the adjusted E. However, if the E value determined this way differs from more than 10% from the expected or handbook value of E for the material being tested, the test data are invalid. In this case the test setup and calibrations must be examined to determine the source of the discrepancy.

Once the crack length for each unloading cycle is known, tabulate the loads (the value of P at the start of each unloading) and crack length compute and plot the K_R, a_{eff} pairs to determine the resistance curve.

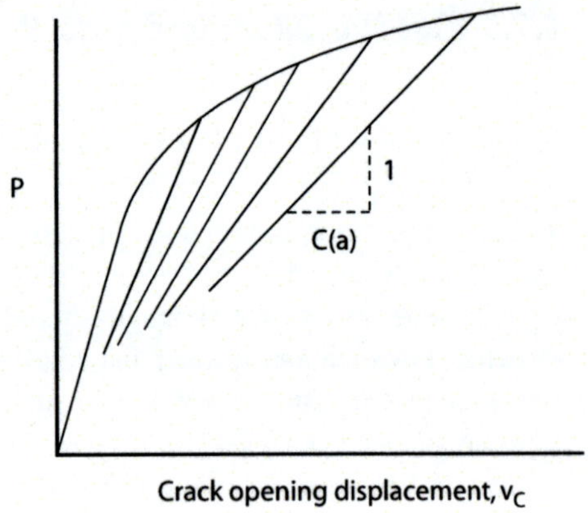

Fig. 6.23 Idealized load vs. crack opening displacement during R-curve testing. Sample is unloaded periodically to determine compliance and hence compute crack length

6.7.3.3 Indirect Approach Using Monotonic Load-Displacement Data

An R curve can also be measured using only a monotonic load-displacement record. In this case, the test is performed slowly, without loading-unloading cycles and the P, v_C data are recorded and plotted. Assuming that the material is elastic, if you were to unload, the unloading curve would follow a line from a particular P, v_C pair back to the origin $(0, 0)$. Thus, for any (P, v_C) the compliance is $C = v_C/P$ and the crack length corresponding to this pair can be determined as can the K_R value. In this way, if the P, v_C record is continuous, a continuous K_R curve can be measured. Although this approach is quite simple a drawback is that there are no checks built into the measurement. In the load-unloading compliance method, the extent to which the unloading curves do not point back to the origin indicates the degree to which the actual behavior deviates from linear elasticity.

A generalization of the R-curve determination method that combines elements of the work of fracture approach with the compliance method is discussed by Sakai and Bradt [31]. Their method addresses the determination of R-curves when the material deviates from linear elastic behavior due to microcracking and other mechanisms.

6.7.4 Sample K–R curve

R curve data for 2024-T351 aluminum are given in [30] and are plotted in Fig. 6.24.

6.8 Exercises

1. In the E399 standard, discuss reasons for measuring the CMOD, v, rather than the load-line displacement (the displacement at the loading pins for the CT or SENB specimens). Discuss reasons for the rule that $P_Q < 1.1 P_{max}$.

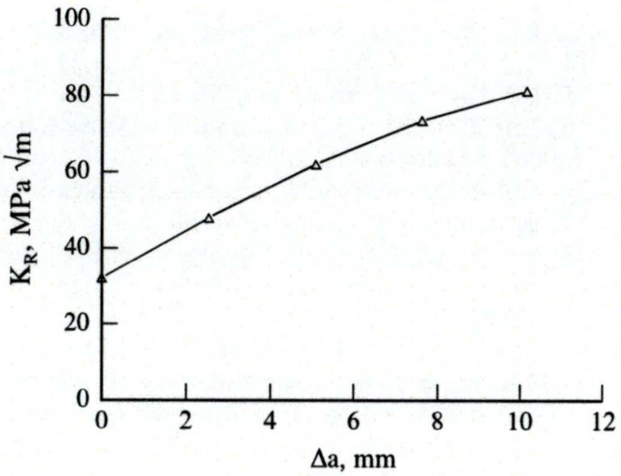

Fig. 6.24 *K–R* data for 2024-T351 aluminum alloy, from [30]

2. Design a test to measure K_{IC} for one of the materials below. Use MKS units for all calculations and dimensions. Include: 1. Specimen geometry, dimensions and mass. 2. Sample calculation showing that constraint conditions would be met. 3. Estimated load and CMOD at fracture (these would be needed to size the testing machine, load cell and clip gauge). 4. Cyclic loading required for fatigue precracking. *Materials*: 1. 7075-T6 aluminum 2. 2024-T3 aluminum 3. ASTM 897 Grade 2 Austempered Ductile Iron 4. Ti-6-4 STA 5. 17-4 PH SS RH 950. Data needed to complete the assignment can be found at www.matweb.com.
3. Why are the conditions $P_{max} < 1.1P_C$ and $-0.05 < \Delta X_0/\Delta X < 0.10$ needed in the chevron-notch test?
4. Discuss reasons why the thickness, B needed for plane-strain constraint in the chevron-notch test is less than that needed for the E-399 test.

References

1. ASTM, *ASTM E399: Standard Test Method for Plain-Strain Fracture Toughness of Metallic Materials* (ASTM International, 2002)
2. ASTM, *ASTM B645: Standard Practice for Linear Elastic Plane-Strain Fracture Toughness of Aluminum Alloys* (ASTM International, 2007)
3. ASTM, *ASTM D 5528: Standard Test Method for Mode I Interlaminar Fracture Toughness of Unidirectional Fiber-Reinforced Polymer Matrix Composites* (ASTM International, 2001)
4. M. Kotaki, M. Hojo, H. Hamada, Z. Maekawa, J. Mater. Sci. Lett. **17**, 515 (1998)
5. M. Kotaki, H. Hamada, Composites, Part A, Appl. Sci. Manuf. **28**, 257 (1997)
6. S. Hashemi, A.J. Kinloch, J.G. Williams, J. Mater. Sci. Lett. **8**, 125 (1989)
7. J.P. Berry, J. Appl. Phys. **34**(1), 62 (1963)
8. R.A. Naik, J.H. Crews, K.N. Shivakumar, *Composite Materials, Fatigue and Fracture*, vol. 3, (1991), p. 169,
9. J.G. Williams, J. Strain Anal. **24**(4), 207 (1989)
10. C.L. Perez, B.D. Davidson, *Proc. 47th AIAA/ASME/ASCE/AHS/ASC Structures, Structural Dynamics and Materials Conference* (2006)
11. B.D. Davidson, X. Sun, J. ASTM Int. **3**(9), JAI1002285 (2006)
12. ASTM, *ASTM D 6671-01: Standard Test Method for Mixed Mode I-Mode II Interlaminar Fracture Toughness of Unidirectional Fiber Reinforced Polymer Matrix Composites* (ASTM International, 2001)

13. B.D. Davidson, V. Sundararaman, Int. J. Fract. **78**, 193 (1996)
14. B.D. Davidson, W. Zhao, J. Compos. Mater. **41**(6), 679 (2007)
15. B.R. Lawn, A.G. Evans, D.B. Marshall, J. Am. Ceram. Soc. **63**, 574 (1980)
16. ASTM, *ASTM E 384: Standard Test Method for Microindentation Hardness of Materials* (ASTM International, 2005)
17. E. Rodeghiero, Synthesis and characterization of metal-ceramic composites produced via sol-gel chemistry. Ph.D. thesis, Cornell University, Ithaca, NY (1998)
18. E. Ustundag, Processing, structure and characterization of in-situ-formed metal-ceramic composites obtained by partial reduction reactions. Ph.D. thesis, Cornell University, Ithaca, NY (1996)
19. D.G. Munz, Int. J. Fract. **16**, R137 (1980)
20. S.X. Wu, in *Chevron-Notched Specimens: Testing and Stress Analysis*. Elastic-Plastic Fracture, vol. 855, ed. by J.H. Underwood, S.W. Freiman, F.I. Baratta (American Society for Testing and Materials, Philadelphia, 1984), pp. 176–192
21. K.R. Brown, F.I. Baratta (eds.), *Chevron-Notch Fracture Test Experience: Metals and Non-metals*. Elastic-Plastic Fracture, vol. 1172 (American Society for Testing and Materials, Philadelphia, 1992)
22. S.M. Barinov, M. Sakai, Int. J. Mater. Res. **9**(6), 1412 (1994)
23. ASTM, *ASTM E 1304: Standard Test Method for Plane-Strain (Chevron-Notch) Fracture Toughness of Metallic Materials* (ASTM International, 2002)
24. A.T. Zehnder, C.Y. Hui, E.D. Rodeghiero, J. Am. Ceram. Soc. **80**(5), 1319 (1997)
25. E.K. Tschegg, Patent 390328, Österreichisches Patentamt (1986)
26. E.K. Tschegg, Patent 396997, Österreichisches Patentamt (1990)
27. E.K. Tschegg, Mater. Prüf. **33**, 338 (1991)
28. E.K. Tschegg, S.E. Stanzl-Tschegg, J. Litzka, in *Reflecting Cracking in Pavements*. Third Rilem Conference, Maastricht, October 2–4 (1996)
29. E.K. Tschegg, M. Elase, S.E. Stanzl-Tschegg, Cem. Concr. Compos. **17**, 57 (1995)
30. ASTM, *ASTM E 561: Standard Practice for K-R Curve Determination* (ASTM International, 2005)
31. M. Sakai, R.C. Bradt, in *Fracture Mechanics of Ceramics*, vol. 7, ed. by R.C. Bradt, A.G. Evans, D.P.H. Hasselman, F.E. Lange (Plenum, New York, 1974), pp. 127–142

Chapter 7
Elastic Plastic Fracture: Crack Tip Fields

Abstract The crack tip stress and strain fields for stationary and growting cracks in elastic-plastic materials are calculated here and the results used to gain insight into ductile fracture criteria such as the R-curve. The size of the plastic zone ahead of the crack is estimated using several models. Knowing the size of the plastic zone relative to the component dimensions is an important step in determining whether to us linear elastic fracture mechanics or elastic-plastic fracture mechanics for a fracture prediction. Effects of large scale yielding are explored through slip line solutions.

7.1 Introduction

Interpreted in the narrowest sense, linear elastic fracture is applicable only to materials that fail in a completely brittle manner. However, virtually all structures or components are made of materials with at least some ductility. Just think of how fragile the man-made world would otherwise be! As discussed, small scale yielding may be invoked to relax the restriction to brittle materials in the many practical applications in which materials undergo small amounts of deformation prior to failure. In this chapter the crack tip fields, energy flows, computational approaches and other considerations of fracture in ductile materials will be discussed. This will provide not only the details of the crack tip fields but will also help us to understand the limitations of the small scale yielding assumption and to understand crack growth resistance curves, temperature dependence of fracture toughness and differences between observed toughness in thin and thick sheets of metal.

The analyses given here will be in the context of small strain, classical plasticity theory. Although not the most general approach to the study of ductile materials, classical plasticity serves as a good starting point for understanding ductile fracture.

7.2 Strip Yield (Dugdale) Model

We will start the discussion not with a problem in elastic plastic fracture but with the strip yield model, an elastic fracture problem. Although this model was developed specifically to model fracture of thin, metal sheets, it has been adapted as a

A.T. Zehnder, *Fracture Mechanics*,
Lecture Notes in Applied and Computational Mechanics 62,
DOI 10.1007/978-94-007-2595-9_7, © Springer Science+Business Media B.V. 2012

prototypical ductile fracture model that yields important scaling relations in elastic plastic fracture and one that illuminates the limitations to small scale yielding.

In the strip yield model a finite crack of length $2a$ is loaded in tension as shown in Fig. 7.1 [1]. It is assumed that the material deforms plastically along very thin zones of length s extending out from the tips of the crack. Inside these zones the stress is limited by the uniaxial yield stress of the material, σ_0, quenching the crack tip stress singularities. This model will apply best to thin sheets of elastic-perfectly plastic materials, i.e. materials that yield with little or no strain hardening. In particular the model is a good approximation for materials exhibiting Lüder's bands. Lüder's bands are regions of locally high deformation that occur in the plastic deformation of materials that contain upper and lower yield points. In such materials, during a uniaxial stress-strain test the stress must first exceed a certain level prior to plastic deformation and then drops down to a steady, lower level as plastic deformation continues.

The problem formulation is a follows: Determine the length s of the yield zones so that the stress is always finite. In doing so we will also determine the stress fields and "crack opening displacement", or COD. The COD is the displacement at $x_1 = \pm a$. It will be shown later that the J-integral for this problem can be expressed as $J = \sigma_0 \delta_T$, where $\delta_T = \text{COD}$.

The boundary conditions for this problem are: As $x_2 \to \infty$: $\sigma_{22} = \sigma_\infty$, $\sigma_{11} = 0$, $\sigma_{12} = 0$. On $x_2 = 0$: $\sigma_{12} = 0$, and on $|x_1| < a$, $\sigma_{22} = 0$, on a $\leq |x_1 + s| < a + s$, $\sigma_{11} = \sigma_0$, and on $|x_1| \geq a + s$, $u_2 = 0$.

A superposition approach can be used for the solution of this problem. The original problem is considered as the superposition of three problems: (i) A crack of length $2(a+s)$ loaded in tension, (ii) a crack of length $2(a+s)$ with closing tractions σ_0 and (iii) a crack with opening tractions over $|x_1| < a$, Fig. 7.1. The goal is to determine the value of s that makes the stress intensity factor of the combined problem equal to zero, hence ensuring that there are no stress singularities in the combined problem, i.e. that the stress is always finite. The total stress intensity factor is $K_I = K_I^{(i)} + K_I^{(ii)} + K_I^{(iii)}$.

The solution to problem (i) of the superposition is $K_I^{(i)} = \sigma_\infty \sqrt{\pi(a+s)}$.

The solution to problem (ii) of the superposition is $K_I^{(ii)} = -\sigma_0 \sqrt{\pi(a+s)}$.

The solution for $K_I^{(iii)}$ is calculated by applying Eq. (5.1) for a crack of length $2(a+s)$ loaded by tractions of $p_2 = \sigma_0$ on $|x_1| < a$.

$$K_I^{(iii)} = \frac{\sigma_0}{\sqrt{\pi(a+s)}} \int_{-a}^{a} \sqrt{\frac{a+s+t}{a+s-t}}\, dt,$$

which integrates to

$$K_I^{(iii)} = 2\sigma_0 \sqrt{\frac{a+s}{\pi}} \tan^{-1} \frac{a}{\sqrt{(a+s)^2 - a^2}}.$$

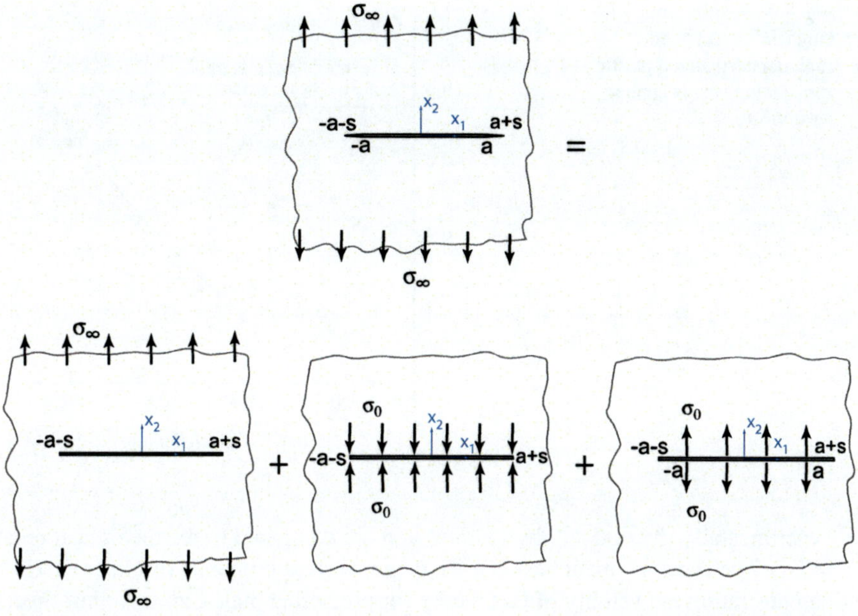

Fig. 7.1 Crack of length $2a$ in infinite plate with tensile loading of $\sigma_{22} = \sigma_\infty$ and strip yield zone with yield stress σ_0 of length s at each crack tip. Yield zone problem can be solved by superposition of the solutions of the three problems sketched

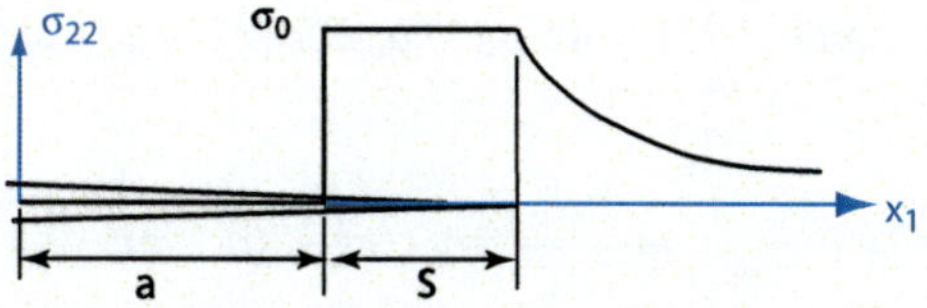

Fig. 7.2 Stress ahead of the crack tip is finite due to strip yield zone

Using the identity $\tan^{-1}\frac{v}{u} = \sin^{-1}\frac{v}{\sqrt{u^2+v^2}}$ and letting $v = a$ and $u = \sqrt{(a+s)^2 - a^2}$ we have

$$K_I^{(iii)} = 2\sigma_0\sqrt{\frac{a+s}{\pi}}\sin^{-1}\frac{a}{a+s}. \tag{7.1}$$

To find the length s of the yield zone we add the stress intensity factors and solve for $K_I = 0$, resulting in

$$s = a\sec\frac{\sigma_\infty \pi}{\sigma_0 2} - a. \tag{7.2}$$

The resulting stress field ahead of the crack, sketched in Fig. 7.2 may be calculated by solving for ϕ' for all three problems and using $\sigma_{22} = \mathrm{Re}\,\phi'$ for $z = x_1$.

Dugdale [1] performed experiments in which sheets of steel containing center and edge cracks were loaded in tension. The plastic zone lengths were measured

Fig. 7.3 Results from Dugdale's experiments showing normalized plastic zone length versus applied tension stress

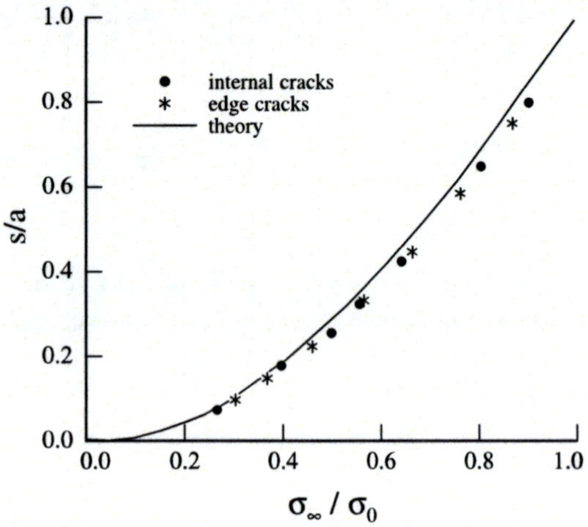

experimentally (by optical observation) and are compared to the theoretical results in Fig. 7.3. The agreement between the theory and experimental results is excellent, demonstrating the validity of this theory for predicting plastic zone in thin sheet of materials with low strain hardening.

The crack opening displacement is also found by superposition. For the combination of problems (i) and (ii), from Eq. (2.81) $u_2^{(i+ii)}(x_1, 0^{(+)}) = (\sigma_\infty - \sigma_0)\frac{\kappa+1}{4\mu}\sqrt{(a+s)^2 - x_1^2}$. For problem (iii) u_2 must be calculated using Eqs. (2.70) and (2.73), with $p_2 = \sigma_0$ for $|x_1| < a$ and $z = x_1$,

$$\phi' = \frac{1}{\pi\sqrt{z^2 - (a+s)^2}} \int_{-a}^{a} \sigma_0 \frac{\sqrt{(a+s)^2 - t^2}}{z - t} dt$$

$$2\mu u_2 = \frac{\kappa+1}{2} \operatorname{Im} \phi$$

$$u_2^{(iii)}(x_1, 0^{(+)}) = \frac{\kappa+1}{4\mu} \frac{\sigma_0}{\pi} \operatorname{Im} \int_{a+s}^{x_1} \frac{1}{\sqrt{z^2 - (a+s)^2}} \int_{-a}^{a} \frac{\sqrt{(a+s)^2 - t^2}}{z - t} dt\, dz. \tag{7.3}$$

The limit of integration is set at $a + s$ so that the crack will be closed, i.e. $u_2 = 0$ at $x_1 = a + s$. Using $s = a \sec \frac{\sigma_\infty \pi}{\sigma_0 2} - a$, integrating the above and putting all the displacements together, for plane stress the crack mouth displacement is [2]

$$\delta_T \equiv u_2(a, 0^{(+)}) - u_2(a, 0^{(-)}) = \frac{8}{\pi} \frac{\sigma_0}{E} a \ln\left[\sec \frac{\sigma_\infty \pi}{\sigma_0 2}\right]. \tag{7.4}$$

.

Although the crack opening profile of the crack is somewhat complicated to calculate for the finite crack, the opening is more easily computed for the case of a

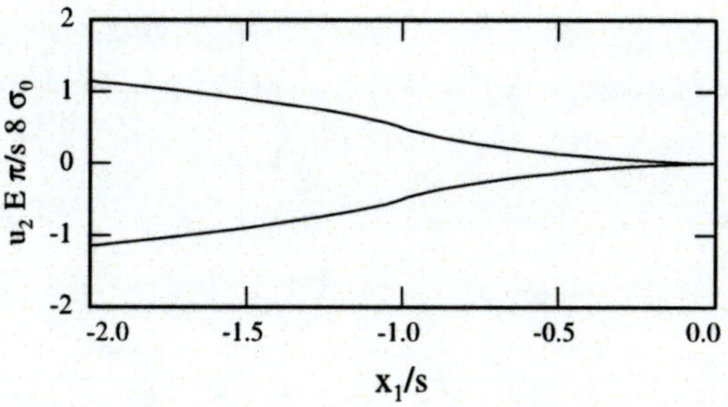

Fig. 7.4 Crack opening profile (normalized) for a Dugdale zone at the tip of a semi-infinite crack under Mode-I loading (see exercise 1). In this case the tip of the yield zone is located at $x_1 = 0$. The crack tip is located at $x_1/s = -1$. Note the cusp-like shape of the opening in the strip yield zone. This shape is also seen in the solution of the finite crack problem

Fig. 7.5 Path Γ for J contour integral for strip yield zone problem

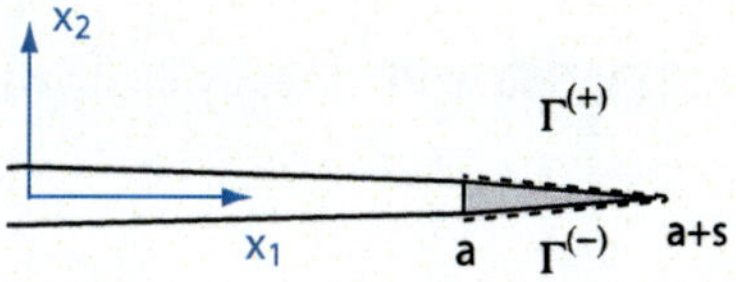

semi-infinite crack under Mode-I loading using Eqs. (5.2) and (5.3), see exercise 1. The resulting crack profile (normalized) is shown in Fig. 7.4. Note the cusp shape of the crack opening in the yield zone.

The J integral may be calculated by shrinking the integration contour, Γ down to the yield zone as shown in Fig. 7.5. Recall that $J = \int_\Gamma (Wn_1 - t_i u_{i,1}) d\Gamma$. Divide the contour into sections $\Gamma^{(+)}$ and $\Gamma^{(-)}$, plus a vertical segment whose length vanishes as the contours are shrunk to the yield zone. On $\Gamma^{(+)}$ $n_1 = 0, n_2 = 1, t_1 = 0, t_2 = \sigma_0$ and $d\Gamma = -dx_1$. On $\Gamma^{(-)}$ $n_1 = 0, n_2 = -1, t_1 = 0, t_2 = -\sigma_0$ and $d\Gamma = dx_1$. Hence the term $t_i u_{i,1}$ in the J integral is $\pm\sigma_0 u_{2,1}$, the Wn_1 term is zero and

$$J = \int_a^{a+s} \sigma_0 u_{2,1}^{(-)}(x_1)\, dx_1 + \int_{a+s}^{a} -\sigma_0 u_{2,1}^{(+)}(x_1)(-dx_1)$$

$$= \sigma_0 \big[u_2^{(-)}(a+s) - u_2^{(-)}(a) + u_2^{(+)}(a) - u_2^{(+)}(a+s) \big].$$

In the above $u_2(a+s) = 0$ and hence $J = \sigma_0(u_2^{(+)}(a) - u_2^{(-)}(a))$. Noting that $u_2^{(+)}(a) = \delta_T/2$ and $u_2^{(-)}(a) = -\delta_T/2$,

$$J = \sigma_0 \delta_T, \tag{7.5}$$

$$J = \frac{8}{\pi} \frac{\sigma_0^2}{E} a \ln \left[\sec \frac{\sigma_\infty \pi}{\sigma_0 2} \right]. \tag{7.6}$$

An important question in the application of linear elastic fracture mechanics is when does small scale yielding (SSY) apply? To explore this question let us compare lengths of the plastic zone and values of the J integral for the full solution and for

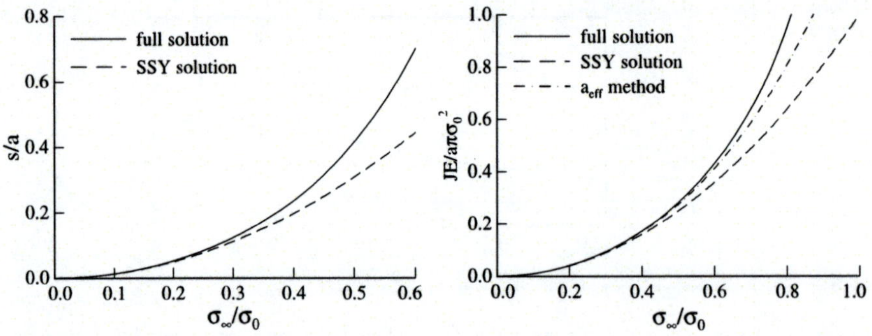

Fig. 7.6 Strip yield zone model. Length of yield zone and values of normalized J integral vs. applied load, σ_∞/σ_0 for a finite crack of length $2a$ in an infinite plate with yield stress σ_0 subject to tension loading σ_∞

SSY. SSY will certainly apply when $s \ll a$. In this case

$$s \approx a\frac{\pi^2}{8}\left(\frac{\sigma_\infty}{\sigma_0}\right)^2, \tag{7.7}$$

$$\delta_T \approx \frac{a\sigma_\infty^2\pi}{E\sigma_0}, \tag{7.8}$$

$$J = \sigma_0\delta_T \approx \frac{a\pi\sigma_\infty}{E}. \tag{7.9}$$

Under conditions of SSY the concept of stress intensity factors still applies, thus using the result that $K_I = \sigma_\infty\sqrt{\pi a}$ we find

$$s = \left(\frac{K_I}{\sigma_0}\right)^2\frac{\pi}{8}, \tag{7.10}$$

$$J = \frac{K_I^2}{E}. \tag{7.11}$$

We see that the relation between J and K_I in the SSY case is the same as in the purely elastic case. Note the scaling of length, s of the yield, or plastic zone. We will observe the same scaling in more sophisticated models of elastic plastic fracture, i.e. that the size of the plastic zone scales as $(K_I/\sigma_0)^2$.

How high could the load be before SSY is no longer valid? Let us plot the plastic zone length, s and J integral value for the full solution and the SSY case. Figure 7.6 shows that up to approximately $\sigma_\infty/\sigma_0 = 0.4$ the SSY and full results are quite close. At this stress level $s/a \approx 0.2$. Thus SSY is valid when the size of the plastic zone is less than approximately 20% of the crack length. More generally SSY is applicable when the plastic zone is a fraction either of the crack length, or, in a finite body, of the remaining ligament length, the distance from the crack tip to the nearest free surface. These estimates can be taken as rules of thumb for the limits of validity of SSY.

Fracture in ductile materials may often have more to do with the accumulation of strain than of the attainment of high stresses. Thus one fracture criterion that could be used would be to postulate that the crack will grow with the crack tip opening displacement (CTOD) exceeds a critical value, i.e. $\delta_T \geq \delta_C$, where δ_T is the CTOD and δ_C is a critical value, representing the maximum accumulation of strains that can occur at the crack tip prior to failure. Since $J = \sigma_0 \delta_T$, it is seen that a critical CTOD fracture criterion is equivalent to a critical J criterion, i.e. $J \geq J_C$.

7.2.1 Effective Crack Length Model

If $J \geq J_C$ is to be used as a fracture criterion, then means for computing J must be available. In cases where extensive plasticity prior to fracture occurs then a full elastic-plastic analysis must be performed. However, in intermediate cases it may be possible to extend LEFM to elastic-plastic fracture through the use of the "effective crack length" concept. In this approach the crack length is artificially increased by a fraction of the plastic zone, i.e. $a_{eff} = a + \eta r_p$, where r_p is the length of the plastic zone (s in the strip yield zone model) and η is an order one factor to be determined. In this approach the effective crack length a_{eff} is substituted for a in the equation for stress intensity factor and J is calculated from

$$J = \frac{K_I^2(a_{eff})}{E}. \tag{7.12}$$

For the problem at hand, $K_I = \sigma_\infty \sqrt{\pi a}$. Thus to determine a_{eff} we solve

$$J = \frac{8}{\pi}\frac{\sigma_0^2}{E} a \ln\left[\sec \frac{\sigma_\infty \pi}{\sigma_0 2}\right] = \frac{K_I^2(a_{eff})}{E} = \frac{\sigma_\infty^2 \pi (a_{eff})}{E}.$$

Expanding the second term in a Taylor series in σ_∞/σ_0 we find that

$$a_{eff} = a + \frac{a\pi^2(\sigma_\infty/\sigma_0)^2}{24}, \tag{7.13}$$

which using the SSY estimate for s can be written as

$$a_{eff} = a + \frac{s}{3}. \tag{7.14}$$

To test the limits of accuracy of the effective crack length concept we plot the effective crack length estimate of J, $J = \frac{\sigma_\infty^2 \pi a}{E}[1 + \frac{\pi^2}{24}(\sigma_\infty/\sigma_0)^2]$ vs. σ_∞/σ_0 in Fig. 7.6. It is seen that using the effective crack length method, the use of LEFM and SSY can be extended in this case to loads of approximately $\sigma_\infty/\sigma_0 \leq 0.6$.

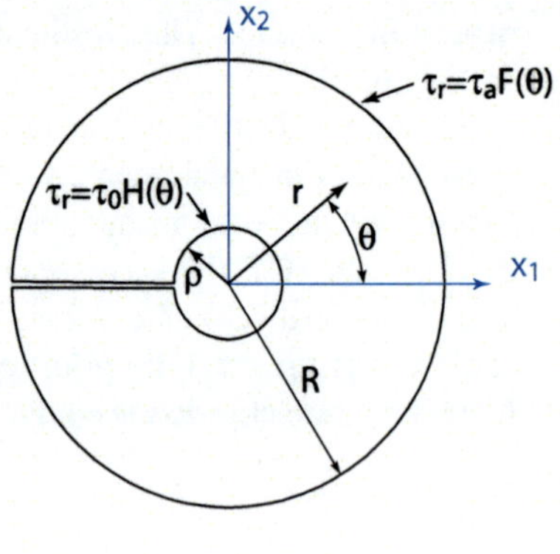

Fig. 7.7 Anti-plane shear crack in circular cylinder with prescribed traction on boundaries. Inner region excludes the crack tip plastic zone, characterized by size ρ. Adapted from [3]

7.3 A Model for Small Scale Yielding

We will consider one more elastic problem before looking at elastic-plastic crack tip fields. In small scale yielding there is an inelastic zone at the crack tip, surrounded by a plastic zone. Hui and Ruina [3] point out that due to crack tip inelasticity the elastic solution does not apply all the way to the crack tip. Thus the elastic solution is valid only for $r > 0$ and the argument that the displacement must be finite as $r \to 0$ used to eliminate the higher order singularities in the Williams expansion no longer applies and hence the higher order singularities must be considered when studying fracture in real materials in which the plastic zone size is non-zero.

The small scale yielding concept depends on two conditions. (1) The K field must be dominant over other terms, both non-singular and more singular, in some annular region around the plastic zone. And (2) specimens which are compared using the SSY assumption must have regions of K field dominance that overlap. These conditions are explored by solving for a model problem in which an anti-plane shear crack inside a cylinder of radius R is assumed to have a crack tip plastic zone that can fit entirely inside a circle of radius ρ at the crack tip [3]. The problem to be solved, Fig. 7.7 has traction free crack faces, prescribed tractions $\tau_a F(\theta)$ on the outer boundary, and prescribed tractions $r = R$, and $\tau_0 H(\theta)$ on the inner boundary, $r = \rho$. The loading on the body is represented by τ_a. From the Dugdale model results we know that ρ, the size of the plastic zone scales as $\rho \sim K^2/\tau_0^2$, where K is the Mode-III stress intensity factor and τ_0 is the flow stress of the material. The stress intensity factor scales as $K_{III} \equiv K \sim \tau_a \sqrt{R}$, since R is the only length scale in the problem and hence serves as a surrogate for the crack length of other characteristic in-plane dimension of a body. The solution for the stress and deformation in the inner region $r < \rho$ is not solved for.

The general solution to the problem shown in Fig. 7.7 is given in Hui and Ruina [3] in terms of an eigenfunction expansion in $z = x_1 + ix_2$. An important result from the general solution is that the stress field includes terms that are *more* singular than $r^{-1/2}$, i.e. terms of order $(\ldots, r^{-5/2}, r^{-3/2})$. Inside the plastic zone the material will have a softer stress-strain response than in the elastic region. The most extreme example would be the case when the material deforms with zero stress. In this case at the tip of the crack there is a hole of radius ρ with $\tau_0 = 0$ as the boundary condition.

Define $\varepsilon = \rho/R$ to the ratio of plastic zone to specimen size. For the case of a hole at the crack tip, $H(\theta) = 0$ the complex stress $\tau = \sigma_{32} + i\sigma_{31}$ is

$$\tau = \tau_a \sum_{n=0}^{\infty} b_n\big[(z/R)^{(2n-1)/2} + \varepsilon^{(2n-1)/2}(\varepsilon R/z)^{(2n+3)/2}\big], \tag{7.15}$$

with

$$b_n = \frac{F_n}{1-\varepsilon^{2n+1}}, \qquad F_n = \frac{1}{\pi}\int_{-\pi}^{\pi} F(\theta)\sin\frac{(2n+1)\theta}{2}\,d\theta. \tag{7.16}$$

The first few terms of this series are

$$\tau = \tau_a\left[b_0\left(\frac{z}{R}\right)^{-1/2} + b_0\varepsilon\left(\frac{z}{R}\right)^{-3/2} + b_1\left(\frac{z}{R}\right)^{1/2} + b_1\varepsilon^3\left(\frac{z}{R}\right)^{-5/2} + \cdots\right].$$

Note that the b_0 term corresponds to the K field (i.e. $1/\sqrt{r}$ term) and that $b_0 = \frac{F_0}{1-\varepsilon}$, i.e. the stress intensity factor can be expressed as

$$K = \frac{K^{applied}}{1-\varepsilon},$$

where $K^{applied} = \sqrt{2\pi R}\tau_a F_0$ is the usual stress intensity factor calculated ignoring the plastic zone. The increase in K due to crack tip plasticity is analogous to the increase in J that occurs due to the yield zone in the Dugdale model. The solution shows that as $\varepsilon = \rho/R \to 0$ the higher order singularity terms go to zero and the stress reverts to the Williams solution, Eq. (2.23).

However, for finite ε (the real world) the higher order terms are important with the $r^{1/2}$, $r^{-1/2}$ and $r^{-3/2}$ terms having the largest coefficients for small ε. A numerical example for $F(\theta) = +1$, $\theta > 0$ and $F(\theta) = -1$, $\theta < 0$ is shown in Fig. 7.8(a) where the normalized stress on $\theta = 0$ is plotted vs r/R and in Fig. 7.8(b) where the K field stress field normalized by the full stress solution is plotted. Near the hole (i.e. near the plastic zone boundary) the $r^{-3/2}$ and $r^{-1/2}$ fields are both important and of comparable magnitude. Further from the hole the K field dominates yielding a minimum fractional error of approximately $\sqrt{\varepsilon}$ at a distance of $r/R \approx \sqrt{\varepsilon}$, or $r \approx \sqrt{\rho R}$, the geometric mean of the plastic zone size and the crack length or other in-plane characteristic dimension. Further from the hole the non-singular terms become as important as the $r^{-1/2}$ term.

The J integral may also be calculated. It is shown that $J = K^2/2\mu$ only when $\varepsilon = 0$. For finite ε, J is increased by two effects. The first is that K increases by $1/(1-\varepsilon)$. The second comes from the interaction of the singular and nonsingular terms. The result is similar to the Dugdale model result where J for finite yield zone length is greater than the J calculated for the elastic case.

Hui and Ruina conclude that (1) The stress field outside the plastic zone under SSY consists of higher order singular terms in addition to those terms given in Eq. (2.23) and that any attempt to improve on the K field should include possible

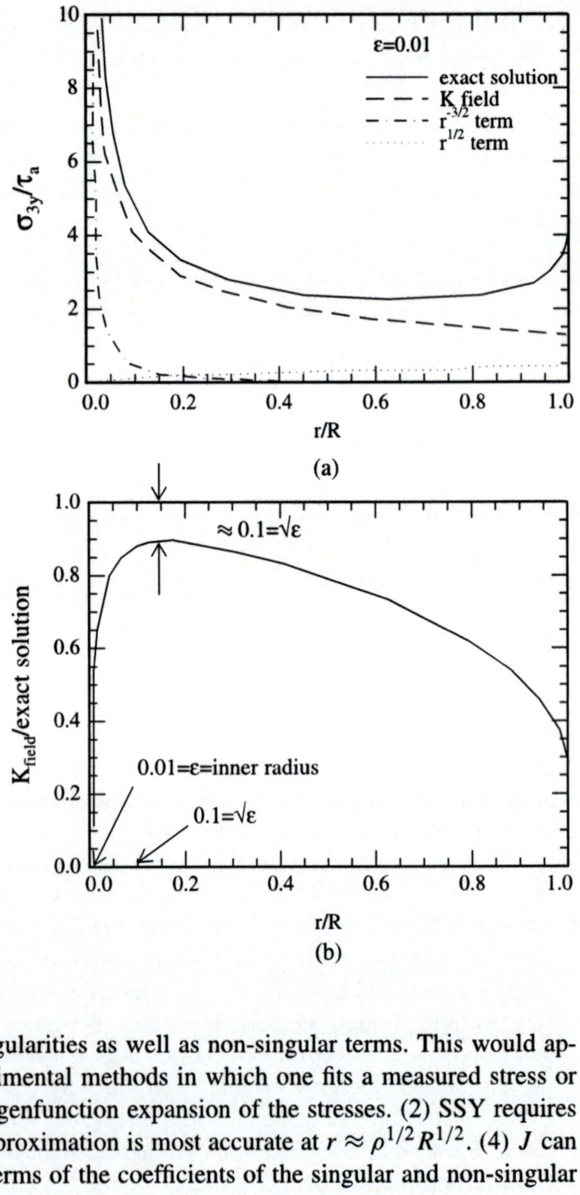

Fig. 7.8 Results from analysis of antiplane shear crack with $\varepsilon = 0.01$ (hole of radius $\rho = 0.01R$ at the crack tip). The applied outer traction is a step function of θ. (**a**) Stress field on $\theta = 0$, comparing full solution to K field solution, $r^{-3/2}$ and $r^{1/2}$ terms. (**b**) Fractional error induced by considering only the K field is minimized at $r \approx \rho^{1/2}R^{1/2}$. Adapted from [3]

effects of higher order singularities as well as non-singular terms. This would apply, for example for experimental methods in which one fits a measured stress or displacement field to an eigenfunction expansion of the stresses. (2) SSY requires $\rho/R \ll 1$. (3) The SSY approximation is most accurate at $r \approx \rho^{1/2}R^{1/2}$. (4) J can be determined exactly in terms of the coefficients of the singular and non-singular terms (if they are known).

7.4 Introduction to Plasticity Theory

Classical plasticity theory [4] assumes that strains are small, ignores the effects of rate of loading and lumps all effects of the history of loading and evolution of microstructure into a simple hardening law. However, despite the limitations of this theory it provides powerful insights into the phenomena of ductile fracture.

Consider the components of stress as forming a six dimensional stress space. A six dimensional strain space can be formed as well with the ij stress and strain components aligned with each other. The yield condition is a surface in stress space and can be written as

$$f(\sigma) = 0. \tag{7.17}$$

If the stresses at a material point lie inside the yield surface, then the deformations are elastic. If the stresses lie on the surface then the deformations may be elastic or elastic-plastic. Stresses outside the yield surface are not allowed; the yield surface will evolve to accommodate attempts to venture beyond it. The ways in which f evolves are quite complex and the subject of much research. The evolution of f is a representation of the evolution of the microstructure of the material as dislocations multiply and dislocation networks interact with each other, typically hardening the material, i.e. increasing the stress needed for further plastic deformation, as deformation accumulates.

As long as strains are small we assume additive decomposition of strains, i.e.

$$\gamma = \gamma^e + \gamma^p, \tag{7.18}$$

where γ^e is the elastic strain and γ^p is the plastic strain. We will assume plastic incompressibility, i.e.

$$\mathrm{Tr}(\gamma^p) = \gamma^p_{kk} = 0. \tag{7.19}$$

The flow rule provides the relation between an increment of strain and an increment of stress. For an elastically isotropic material, we use the associative flow rule in which increments of plastic strain are orthogonal to the yield surface,

$$d\gamma_{ij} = \frac{1+\nu}{E} d\sigma_{ij} - \frac{\nu}{E} d\sigma_{kk}\delta_{ij} + \frac{1}{H}\langle \mathbf{n} \cdot d\sigma \rangle n_{ij}, \tag{7.20}$$

where $\mathbf{n}$ is a vector in stress space that is normal to the yield surface, f,

$$\mathbf{n} = \frac{\partial f/\partial \sigma}{[\partial f/\partial \sigma \cdot \partial f/\partial \sigma]^{1/2}}, \tag{7.21}$$

H is an inverse stress-strain slope, and the brackets denote a function such that $\langle x \rangle = x$ for $x > 0$, $\langle x \rangle = 0$ for $x < 0$.

You are most likely familiar with the Tresca and von-Mises yield criteria. Here we will work with the von-Mises theory with isotropic hardening,

$$f = J_2 - \frac{\sigma_0^2}{3}, \tag{7.22}$$

where σ_0 is the uniaxial yield stress. The yield stress will evolve in response to deformation. Note that this theory does not include important phenomena such as the Bauschinger effect in which the tensile and compressive yield stresses differ.

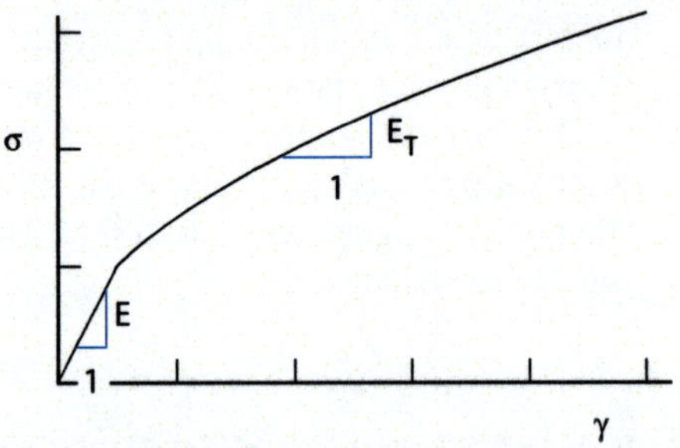

Fig. 7.9 Schematic stress-strain curve showing elastic modulus prior to yield and tangent modulus after yield

We define equivalent stress, σ_e, and equivalent plastic strain γ_e^p so that plots of σ vs γ^p in a uniaxial test and plots of σ_e vs γ_e^p are the same,

$$\sigma_e = \sqrt{3J_2}, \tag{7.23}$$

$$d\gamma_e^p = \sqrt{\frac{2}{3} d\gamma^p \cdot d\gamma^p}, \tag{7.24}$$

where J_2 is the second invariant of the deviatoric stress,

$$J_2 = \frac{1}{2}\mathbf{S} \cdot \mathbf{S} = \frac{1}{2} S_{ij} S_{ij}, \tag{7.25}$$

$$\mathbf{S} = \sigma - \frac{1}{3}\operatorname{Tr}(\sigma)\mathbf{I}, \text{ or } S_{ij} = \sigma_{ij} - \frac{1}{3}\sigma_{kk}\delta_{ij}. \tag{7.26}$$

In this case the yield function can be written as $\sigma_e^2 - \sigma_0^2 = 0$ and

$$\frac{1}{H} = \frac{3}{2}\left(\frac{1}{E_T} - \frac{1}{E}\right), \tag{7.27}$$

where E_T is the tangent modulus, the instantaneous slope of the uniaxial stress-strain curve, see Fig. 7.9.

With some work you can show that $\frac{\partial J_2}{\partial \sigma_{ij}} = S_{ij}$. Using this result and noting that $\frac{\partial f}{\partial \sigma_{ij}} = \frac{\partial J_2}{\partial \sigma_{ij}}$, Eq. (7.21) for $\mathbf{n}$ gives the direction of plastic strain increment as

$$n_{ij} = \frac{S_{ij}}{\sqrt{2J_2}}. \tag{7.28}$$

Again, with some work you can show that the increment of plastic strain is given by

$$d\gamma_{ij}^p = \frac{1}{H}\langle \mathbf{n} \cdot d\sigma \rangle n_{ij} = \frac{1}{H}\frac{\langle dJ_2 \rangle}{2J_2} S_{ij}. \tag{7.29}$$

Note that $dJ_2 = S_{mn}\, dS_{mn}$.

In the incremental, or "flow" theory of plasticity we deal with only with increments of plastic strain, preserving the differences between loading and unloading behavior. During loading, the stress follows the nonlinear curve shown in

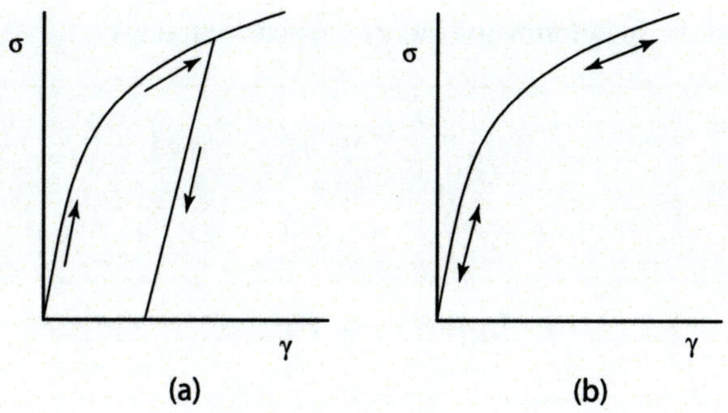

Fig. 7.10 (**a**) Incremental plasticity theory. (**b**) Deformation plasticity theory

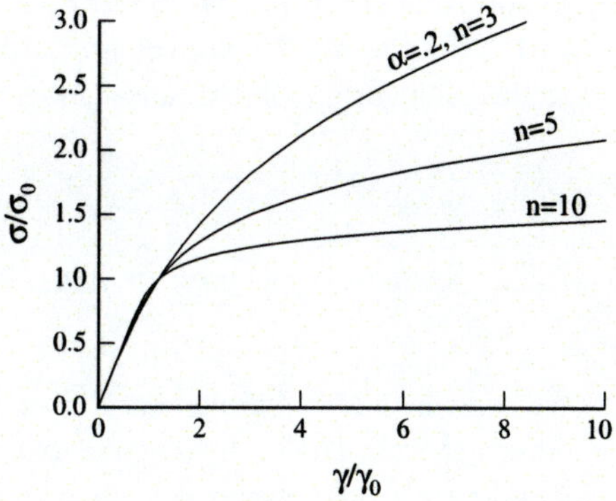

Fig. 7.11 Uniaxial stress strain relation for power-law hardening model. In this model the yield stress evolves as a unique function of plastic strain

Fig. 7.10(a). During unloading the deformation is linearly elastic. The deformation theory of plasticity is not a plasticity theory at all, but rather a nonlinear elasticity theory in which the loading and unloading behavior are the same, Fig. 7.10(b). The utility of deformation theory is that as long as the loading is proportional, i.e. at a material point $\sigma_{ij}(\mathbf{x},t) = \eta(\mathbf{x},t)\hat{\sigma}_{ij}$, then the incremental and deformation theories agree. If proportional loading is not maintained, then incremental and deformation theory results will differ.

For fracture problems it is convenient to consider the power law model of hardening in which the uniaxial stress-strain curve for loading is given by

$$\gamma^p = \gamma_0 \alpha \left(\frac{\sigma}{\sigma_0} \right)^n, \tag{7.30}$$

where σ_0 is the initial yield stress, α and n are power law fitting parameters and γ_0 is the strain at first yield, i.e. $\gamma_0 = \sigma_0 / E$. Higher values of n correspond to lower amounts of strain hardening, see Fig. 7.11 The nonhardening, elastic-perfectly plastic limit is reached when $n \to \infty$. In this case the yield stress, σ_0 is constant.

In incremental theory the plastic strain increment is

$$d\gamma_{ij}^p = \frac{3}{2} n \gamma_0 \alpha \left(\frac{\sigma_e}{\sigma_0} \right)^{n-2} \frac{S_{ij}}{\sigma_0} \frac{\langle d\sigma_e \rangle}{\sigma_0}. \tag{7.31}$$

In deformation theory the plastic strain is

$$\gamma_{ij}^{p} = \frac{3}{2}\gamma_0\alpha\left(\frac{\sigma_e}{\sigma_0}\right)^{n-1}\frac{S_{ij}}{\sigma_0}. \tag{7.32}$$

7.5 Anti-plane Shear Cracks in Elastic-Plastic Materials in SSY

The relative simplicity of anti-plane shear theory will allow us to find complete solutions for crack tip fields under SSY conditions. These solutions provide a great deal of insight into the more complex Mode-I and Mode-II problems and provide a hint as to the origin of resistance curves in the fracture of ductile materials.

7.5.1 *Stationary Crack in Elastic-Perfectly Plastic Material*

We consider a semi-infinite crack in an elastic-perfectly plastic material under conditions of SSY. The far field boundary conditions are that as $r \to \infty$ the stress field must revert to the elastic, K_{III} field, Eq. (2.23). As shown in Fig. 7.12 there is a "plastic zone" ahead of the crack tip. Inside this zone the yield criterion is met and the strain field has an elastic and a plastic component. Outside this zone the strains are purely elastic. We will show that the plastic zone is circular with radius R and that the stresses inside the plastic zone can be matched to the elastic filed by shifting the origin of the elastic field ahead by $R/2$. This is similar to the idea of effective crack length introduced in Sect. 7.2.1.

To simplify the notation, let $\tau_r = \sigma_{r3}$, $\tau_\theta = \sigma_{\theta 3}$, $\gamma_r = 2\gamma_{r3}$, $\gamma_\theta = 2\gamma_{\theta 3}$, $K = K_{III}$ and $w = u_3(r,\theta)$.

The equilibrium equation in polar coordinates is

$$\frac{1}{r}\frac{\partial}{\partial r}(r\tau_r) + \frac{1}{r}\frac{\partial \tau_\theta}{\partial \theta} = 0. \tag{7.33}$$

Inside the plastic zone the yield condition must be satisfied. In anti-plane shear the yield condition simplifies to

$$\tau_r^2 + \tau_\theta^2 = \tau_0^2, \tag{7.34}$$

where τ_0 is the yield stress in shear, i.e. the yield surface is a circle of radius τ_0 in τ_r, τ_θ space. By the normality rule the plastic strain increments are $d\gamma_r = d\lambda^p\tau_r$, $d\gamma_\theta = d\lambda^p\tau_\theta$. If at a point the stress state remains at the same location on the yield surface, then the total plastic strain will be $\gamma_r^p = \lambda^p\tau_r$ and $\gamma_\theta = \lambda^p\tau_\theta$. The elastic

Fig. 7.12 Plastic zone and coordinate geometry for solution of stationary anti-plane shear crack in elastic-perfectly plastic material

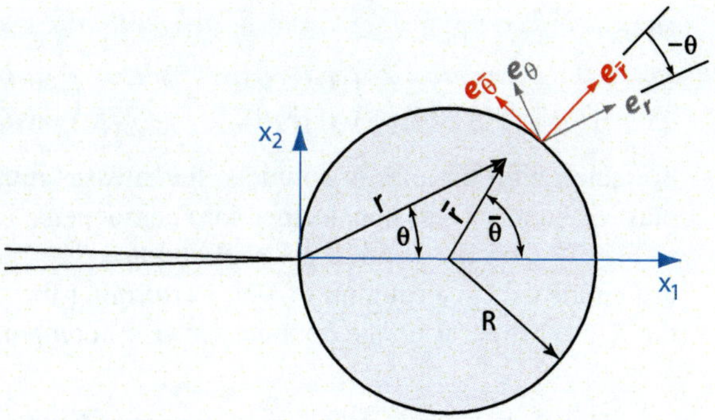

strains are always $\gamma = \tau/\mu$, thus in the plastic zone, the constitutive relation is

$$\begin{aligned} \gamma_r &= \lambda \tau_r, \\ \gamma_\theta &= \lambda \tau_\theta, \\ \lambda &= \frac{1}{\mu} + \lambda^p. \end{aligned} \tag{7.35}$$

The strain displacement relations are

$$\begin{aligned} \gamma_r &= \frac{\partial w}{\partial r}, \\ \gamma_\theta &= \frac{1}{r}\frac{\partial w}{\partial \theta}. \end{aligned} \tag{7.36}$$

Following [2] the solution given in [5] is given briefly. Consider a solution where the plastic zone is a circle of radius R as shown in Fig. 7.12, the stresses inside the plastic zone are given by $\tau_r = 0$ and $\tau_\theta = \tau_0$ and the stresses outside the plastic zone are given by the elastic K_{III} field offset by a distance R ahead of the crack tip.

It is easily seen that the equilibrium equation and yield condition are both satisfied. From Eq. (7.35) $\gamma_r = 0$ and $\gamma_\theta = \lambda(r,\theta)\tau_0$. From the strain-displacement relation, $\partial w/\partial r = 0$ which can be integrated to yield $w = w(\theta) = F(\theta)$, where F is an as yet unknown function. Using $\gamma_\theta = \frac{1}{r}\frac{\partial w}{\partial \theta} = \lambda(r,\theta)\tau_0$ we find that

$$\lambda(r,\theta) = \frac{F'(\theta)}{r\tau_0}. \tag{7.37}$$

To find F let us match the traction and displacement of the plastic solution to those of the shifted K_{III} field at the elastic-plastic boundary. From Eq. (2.23) in terms of $\bar{r}$ and $\bar{\theta}$ coordinates the elastic field is

$$w(\bar{r},\bar{\theta}) = \sqrt{\frac{2}{\pi}}\frac{K}{\mu}\bar{r}^{1/2}\sin\bar{\theta}/2$$

and

$$\begin{pmatrix} \tau_{\bar{r}}(\bar{r},\bar{\theta}) \\ \tau_{\bar{\theta}}(\bar{r},\bar{\theta}) \end{pmatrix} = \frac{K}{\sqrt{2\pi \bar{r}}} \begin{pmatrix} \sin\bar{\theta}/2 \\ \cos\bar{\theta}/2 \end{pmatrix}.$$

To match with the plastic solutions the stresses must be evaluated on the elastic-plastic boundary and transformed into components in the (r,θ) system. As defined in Fig. 7.12 on the boundary $\bar{r} = R$ and $\bar{\theta} = 2\theta$. The coordinate transformation is accomplished by a rotation of $-\theta$. Performing the evaluation and transformation, the K field stresses on the boundary in (r,θ) components are

$$\tau_r = \tau_{\bar{r}} \cos\theta - \tau_{\bar{\theta}} \sin\theta = \frac{K}{\sqrt{2\pi R}} (\sin\theta\cos\theta - \sin\theta\cos\theta) = 0,$$

$$\tau_\theta = \tau_{\bar{r}} \sin\theta + \tau_{\bar{\theta}} \cos\theta = \frac{K}{\sqrt{2\pi R}}.$$

Since $\tau_r = 0$ both inside and outside the plastic zone we need only to match τ_θ to match the tractions across the plastic zone boundary, i.e. set $\tau_\theta = \frac{K}{\sqrt{2\pi R}} = \tau_0$. Solving for R, the radius of the plastic zone,

$$R = \frac{1}{2\pi}\left(\frac{K}{\tau_0}\right)^2. \tag{7.38}$$

Matching the displacement field on $\bar{r} = R$ and $\bar{\theta} = 2\theta$,

$$w(R,\theta) = \sqrt{\frac{2}{\pi}}\frac{K}{\mu} R^{1/2} \sin\theta = F(\theta),$$

or

$$F(\theta) = 2R\gamma_0 \sin\theta, \tag{7.39}$$

where $\gamma_0 = \tau_0/\mu$ is the yield strain.

The strain field in the plastic zone is found from Eq. (7.37)

$$\begin{aligned} \lambda(r,\theta) &= \frac{F'(\theta)}{r\tau_0} = \frac{2R\cos\theta}{\mu r}, \\ \gamma_\theta(r,\theta) &= \lambda(r,\theta)\tau_0 = \frac{2\gamma_0 R\cos\theta}{r}. \end{aligned} \tag{7.40}$$

To summarize: The plastic zone size is

$$r_p = 2R = \frac{1}{\pi}\left(\frac{K}{\tau_0}\right)^2. \tag{7.41}$$

Inside the plastic zone the fields are

$$w = \frac{K}{\mu}\sqrt{\frac{2R}{\pi}} \sin\theta,$$

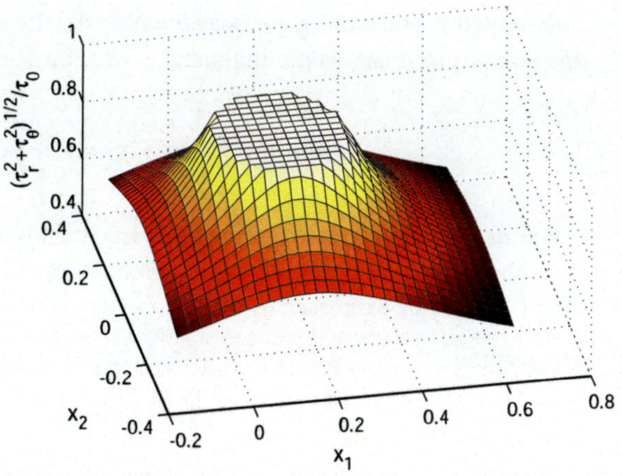

Fig. 7.13 Stress field $\sqrt{\tau_r^2+\tau_\theta^2}$ for anti-plane shear crack in elastic-perfectly plastic material

$$\begin{aligned}\tau_r &= 0,\\ \tau_\theta &= \tau_0,\\ \gamma_r &= 0,\\ \gamma_\theta &= \frac{2\gamma_0 R}{r}\cos\theta,\end{aligned}$$

where $\gamma_0 = \frac{\tau_0}{\mu}$. *Outside* the plastic zone

$$\begin{aligned}w &= \frac{K}{\mu}\sqrt{\frac{2\bar{r}}{\pi}}\sin\bar{\theta}/2,\\ \tau_r &= \tau_{\bar{r}}\cos\theta - \tau_{\bar{\theta}}\sin\theta,\\ \tau_\theta &= \tau_{\bar{r}}\sin\theta + \tau_{\bar{\theta}}\cos\theta,\\ \tau_{\bar{r}} &= \frac{K}{\sqrt{2\pi\bar{r}}}\sin\bar{\theta}/2,\\ \tau_{\bar{\theta}} &= \frac{K}{\sqrt{2\pi\bar{r}}}\cos\bar{\theta}/2,\end{aligned}$$

where at a given (x_1, x_2), $r = \sqrt{x_1^2 + x_2^2}$, $\bar{r} = \sqrt{(x_1 - R)^2 + x_2^2}$, $\theta = \tan^{-1} x_2/x_1$, and $\bar{\theta} = \tan^{-1} x_2/(x_1 - R)$.

Note that the size R of the plastic zone scales as $(K/\tau_0)^2$ exactly as with the Dugdale zone. The strain field is singular as $1/r$, in contrast to the $1/\sqrt{r}$ singularity for the elastic field. The plastic deformation allows a stronger accumulation of strain at the crack tip. However the stress is bounded, in contrast to the $1/\sqrt{r}$ singularity of the elastic field. As an illustration, the values of $\sqrt{\tau_r^2 + \tau_\theta^2}/\tau_0$ are plotted in Fig. 7.13.

Noting that at the crack surfaces $\theta = \pm\pi$ the deformation is elastic, the crack tip opening displacement is (noting that $J = K^2/2\mu$ for SSY)

$$\delta_T \equiv w(\pi) - w(-\pi) = \frac{2K^2}{\pi\mu\tau_0} = \frac{J}{\tau_0}\frac{4}{\pi}. \tag{7.42}$$

Note that the scaling of the crack tip displacement is J/τ_0, the same scaling obtained from the Dugdale model, with τ_0 replacing σ_0. In terms of J, the plastic zone, Eq. (7.41) can be expressed as

$$r_p = \frac{2J}{\pi\tau_0\gamma_0}. \tag{7.43}$$

Note that since the stress and the strain distribution are constant in the plastic zone (with the strain magnitude increasing in proportion to the plastic zone size, $2R$) proportional loading is attained and thus deformation theory of plasticity can be applied up to the point that the crack is unloaded, or begins to grow. Deformation theory is a nonlinear elasticity theory, thus the J integral is path independent.

7.5.2 Stationary Crack in Power-Law Hardening Material

For a power-law hardening material the results are similar [5]. The plastic zone is a circle of radius R but its center is not at $x_1 = R$, but rather at $x_1 = \frac{n-1}{n+1}R$, thus the plastic zone extends ahead of and behind the crack tip. The plastic zone size is

$$r_p = \frac{n}{\pi(n+1)}\left(\frac{K_{III}}{\tau_0}\right)^2. \tag{7.44}$$

Note that the plastic zone size decreases with decreasing n, i.e. more hardening.

Inside the plastic zone the stress and strain are given by

$$\tau_\alpha = \tau_0\left(\frac{r_p}{r}\right)^{1/(n+1)}\tilde{\tau}_\alpha(\theta), \tag{7.45}$$

$$\gamma_\alpha = \gamma_0\left(\frac{r_p}{r}\right)^{n/(n+1)}\tilde{\gamma}_\alpha(\theta). \tag{7.46}$$

Outside the plastic zone the stresses are given by

$$\tau = \frac{K_{III}}{\sqrt{2\pi(z - \frac{n-1}{n+1}R)}}.$$

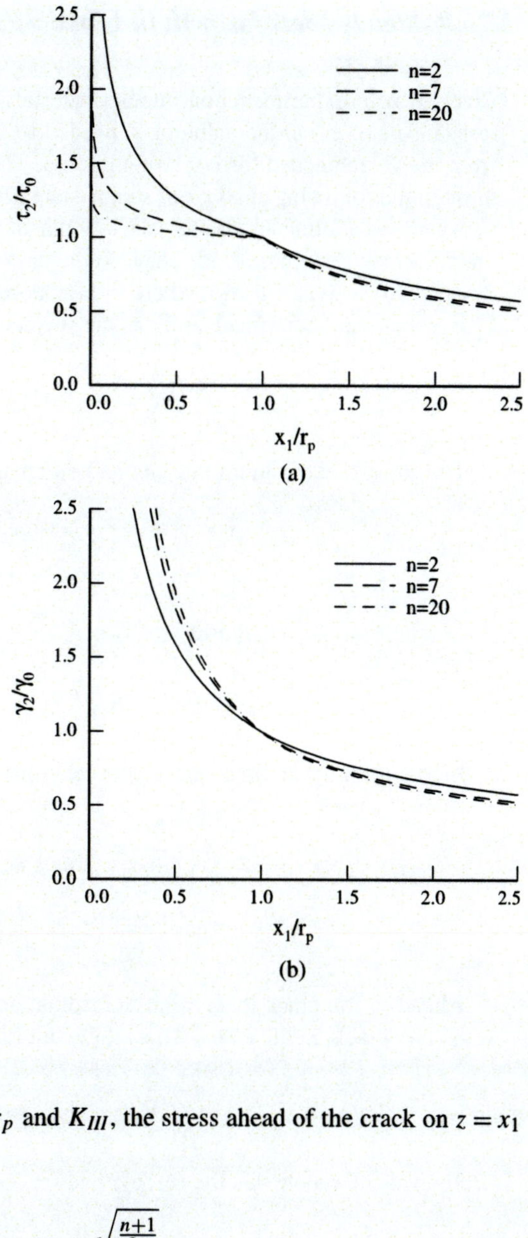

Fig. 7.14 (**a**) Stress field on $\theta = 0$ inside and outside plastic zone for power law hardening materials. (**b**) Strain field

Using the relations between R, r_p and K_{III}, the stress ahead of the crack on $z = x_1$ may be written as

$$\tau_2 = \tau_0 \frac{\sqrt{\frac{n+1}{2n}}}{\sqrt{\frac{x_1}{r_p} - \frac{n-1}{2n}}}. \tag{7.47}$$

The stress and strain inside and outside of the plastic zone are plotted in Figs. 7.14(a) and 7.14(b). As $n \to \infty$ the stress inside the plastic zone is constant and $\gamma \sim r^{-1}$.

7.5.3 Steady State Growth in Elastic-Perfectly Plastic Material

Crack growth experiments on ductile materials generally show a steeply rising J resistance curve, i.e. the value of J needed to continue to grow a crack increases from the value needed for fracture initiation. This can be understood at some level through the following analysis of steady state crack growth.

Again we assume an elastic-perfectly plastic material. Further more we assume that the crack growth is steady state, so that any other field quantity, e.g. g, is given by $g(x_1, x_2) = g(x_1' - a, x_2)$, where x_1 is a coordinate that moves with the crack tip, x_1' is a stationary coordinate and a is the current crack length. In this case

$$\frac{\partial g}{\partial a} \equiv \dot{g} = \frac{\partial g}{\partial x_1}(-1) = -g_{,1}\,.$$

The stress and the strain rate can be written as vectors,

$$\tau = \tau_1 \mathbf{e}_1 + \tau_2 \mathbf{e}_2 = \tau_r \mathbf{e}_r + \tau_\theta \mathbf{e}_\theta,$$

$$\dot{\gamma} = \dot{\gamma}_1 \mathbf{e}_1 + \dot{\gamma}_2 \mathbf{e}_2 = \dot{\gamma}_r \mathbf{e}_r + \dot{\gamma}_\theta \mathbf{e}_\theta.$$

In rate form the constitutive law is

$$\dot{\gamma} = \dot{\gamma}^e + \dot{\gamma}^p = \frac{\dot{\tau}}{\mu} + \dot{\lambda}^p \tau.$$

In rate form the strain displacement relations are

$$\begin{aligned} \dot{\gamma}_r &= \frac{\partial \dot{w}}{\partial r} = -\frac{\partial}{\partial r}\left(\frac{\partial w}{\partial x_1}\right), \\ \dot{\gamma}_\theta &= \frac{1}{r}\frac{\partial \dot{w}}{\partial \theta} = -\frac{1}{r}\frac{\partial}{\partial \theta}\left(\frac{\partial w}{\partial x_1}\right). \end{aligned} \tag{7.48}$$

Ahead of the crack there is no unloading, so the stress state is the same as the stationary crack, i.e. $\tau = 0\mathbf{e}_r + \tau_0 \mathbf{e}_\theta$. Substituting into the constitutive law

$$\dot{\gamma} = \frac{\dot{\tau}}{\mu} + \dot{\lambda}^p \tau_0 \mathbf{e}_\theta.$$

To evaluate $\dot{\tau}$ note that the $(\mathbf{e}_1, \mathbf{e}_2)$ and $(\mathbf{e}_r, \mathbf{e}_\theta)$ basis vectors are related by

$$\mathbf{e}_\theta = -\sin\theta \mathbf{e}_1 + \cos\theta \mathbf{e}_2,$$

$$\mathbf{e}_r = \cos\theta \mathbf{e}_1 + \sin\theta \mathbf{e}_2.$$

Thus

$$\frac{\partial \mathbf{e}_\theta}{\partial x_1} = -\cos\theta \frac{\partial \theta}{\partial x_1}\mathbf{e}_1 - \sin\theta \frac{\partial \theta}{\partial x_1}\mathbf{e}_2,$$

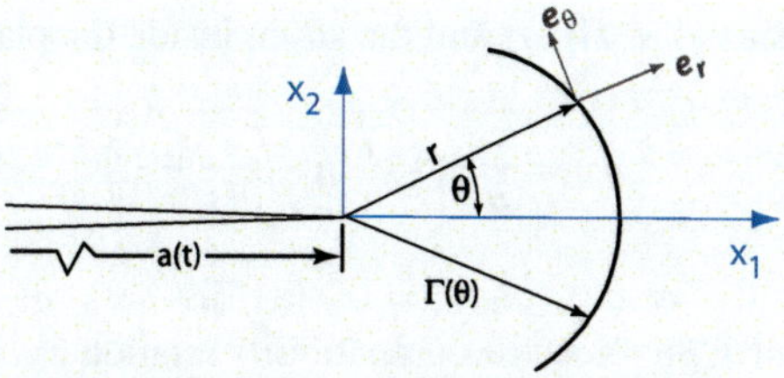

Fig. 7.15 Growing anti-plane shear crack in elastic-perfectly plastic material. $\Gamma(0) \equiv r_p$. Finite element results [6] show that the size of the plastic zone along $x_2 = 0$ is approximately the same as for the stationary crack and that the plastic zone for the moving crack is roughly wedge shaped with an angle of approximately ±20° from the crack tip

$$\frac{\partial \theta}{\partial x_1} = \frac{\partial}{\partial x_1} \tan^{-1} \frac{x_2}{x_1} = -\frac{\sin \theta}{r},$$

$$\dot{\mathbf{e}}_\theta = -\frac{\partial \mathbf{e}_\theta}{\partial x_1} = -\frac{\sin \theta}{r} \mathbf{e}_r$$

$$\Rightarrow \quad \dot{\tau} = \tau_0 \dot{\mathbf{e}}_\theta = -\frac{\tau_0 \sin \theta}{r} \mathbf{e}_r.$$

The strain rate can now be written as

$$\dot{\gamma} = -\frac{\tau_0}{\mu} \frac{\sin \theta}{r} \mathbf{e}_r - \dot{\lambda}^p \tau_0 \mathbf{e}_\theta. \tag{7.49}$$

Substituting Eq. (7.49) into the strain-rate displacement relations Eq. (7.48) we have

$$\dot{\gamma}_r = -\frac{\partial}{\partial r}\left(\frac{\partial w}{\partial x_1}\right) = \frac{-\tau_0 \sin \theta}{\mu r}, \tag{7.50}$$

$$\dot{\gamma}_\theta = -\frac{1}{r} \frac{\partial}{\partial \theta}\left(\frac{\partial w}{\partial x_1}\right) = \dot{\lambda}^p \tau_0. \tag{7.51}$$

Integrating Eq. (7.50) with respect to r we have

$$\frac{\partial w}{\partial x_1} \equiv \gamma_1 = \frac{\tau_0}{\mu} \sin \theta \ln r + F(\theta).$$

The shape and extent of the elastic-plastic boundary are not known, but let the boundary be denoted by $\Gamma(\theta)$ as shown in Fig. 7.15. Along this boundary the elastic stress-strain relation must be satisfied, i.e. $\gamma = (\tau_0/\mu)\mathbf{e}_\theta = (\tau_0/\mu)(-\sin \theta \mathbf{e}_1 + \cos \theta \mathbf{e}_2)$. Matching the γ_1 components of strain from the elastic and elastic-plastic solutions,

$$\gamma_1 = -\frac{\tau}{\mu} \sin \theta = \frac{\tau_0}{\mu} \sin \theta \ln \Gamma(\theta) + F(\theta),$$

$F(\theta) = -(\tau_0/\mu)\sin\theta(1 + \Gamma(\theta))$ and the strain inside the plastic zone is

$$\gamma_1 = \frac{\tau_0}{\mu}\sin\theta\left(\ln\frac{r}{\Gamma(\theta)} - 1\right). \tag{7.52}$$

Now consider the results on $\theta = 0$. In this case $\partial(\cdot)/\partial x_1 = \partial(\cdot)/\partial r$ and $\partial(\cdot)/\partial x_2 = (1/r)\partial(\cdot)/\partial\theta$. Note the compatibility relation $\partial\gamma_1/\partial x_2 = \partial\gamma_2/\partial x_1$. Thus

$$\frac{\partial\gamma_1}{\partial x_2} = \frac{1}{r}\frac{\partial\gamma_1}{\partial\theta} = \frac{\partial\gamma_2}{\partial x_1} = \frac{\partial\gamma_\theta}{\partial r}.$$

Thus

$$\begin{aligned}\frac{\partial\gamma_\theta}{\partial r} &= \frac{1}{r}\frac{\partial}{\partial\theta}\left(\frac{\tau_0}{\mu}\sin\theta\left(\ln\frac{r}{\Gamma(\theta)} - 1\right)\right),\\ &= \frac{1}{r}\frac{\tau_0}{\mu}\left(\ln\frac{r}{\Gamma(0)} - 1\right) \quad \text{(setting } \theta = 0 \text{ after differentiating).}\end{aligned} \tag{7.53}$$

Denote $\Gamma(0) = r_p$, the size of the plastic zone ahead of the crack. Integrate Eq. (7.53) with respect to r with the condition that $\gamma_\theta = \tau_0/\mu = \gamma_0$ on $r = r_p$ to get

$$\gamma_\theta(r, 0) = \frac{\tau_0}{\mu}\left(\frac{1}{2}\ln^2\frac{r}{r_p} - \ln\frac{r}{r_p} + 1\right). \tag{7.54}$$

Contrast this with the strain field for a stationary crack

$$\gamma_\theta(r, 0) = \frac{r_p}{r}\frac{\tau_0}{\mu}.$$

The weaker log singularity of the moving crack implies that for the same plastic zone size the moving crack has lower strain than the moving crack. This is shown in Fig. 7.16(a) where the strains for the moving and stationary crack are plotted.

The weaker strain singularity for the moving crack has important implications for the resistance curve. Let us assume a simple criterion for fracture, namely that the crack will grow the strain reaches a critical value at a fixed distance ahead of the crack tip, i.e. $\gamma = \gamma_C$ at $r = r_C$. Let us also assume small scale yielding so that $J = K^2/2\mu$. Finite element simulations of crack growth in anti-plane shear [6] show that the plastic zone size is approximately the same for steady-state growth as for stationary cracks, thus let us assume that the relation of plastic zone length to J is given by the stationary crack result, Eq. (7.43). For the *stationary* crack

$$\begin{aligned}\frac{\gamma_C}{\gamma_0} &= \frac{r_p}{r_C}\\ &= \frac{2J_C}{\pi\tau_0\gamma_0}\frac{1}{r_C},\end{aligned}$$

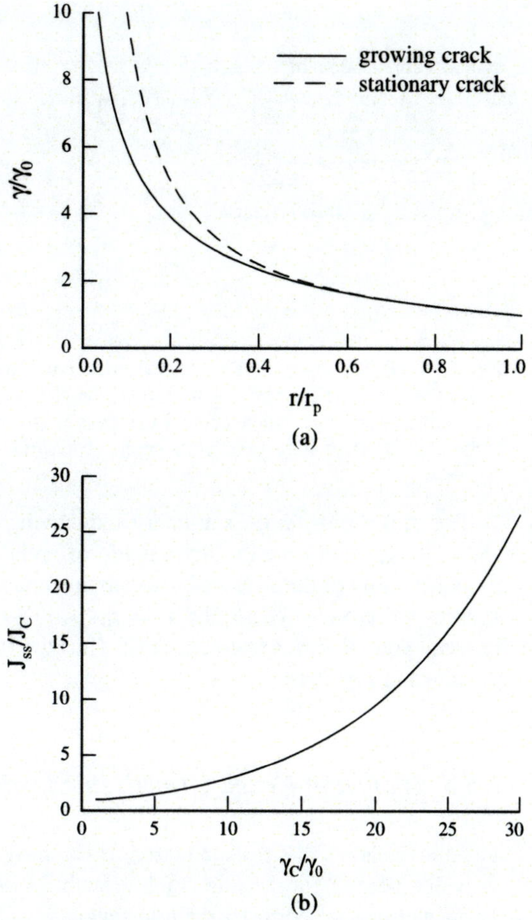

Fig. 7.16 (**a**) Strain distribution ahead of stationary and steady state moving, antiplane shear crack in elastic-perfectly plastic materials. (**b**) J_{ss}/J_C is the value of J needed to drive a steady state crack, normalized by value of J needed to initiate crack growth. Calculated based on critical strain, γ_C at a fixed distance ahead of the crack

where J_C is the value of J needed to satisfy the failure condition. Solving for J_C:

$$J_C = \frac{\pi r_c \tau_0 \gamma_c}{2}. \tag{7.55}$$

For the *moving crack* let J_{ss} be the steady state value of J needed to drive the crack. Setting $\gamma = \gamma_C$ in Eq. (7.54) and $J = J_{ss}$ in Eq. (7.43), and solving

$$J_{ss} = \frac{\pi r_C \tau_0 \gamma_0}{2} \exp\left(\sqrt{2\gamma_C/\gamma_0 - 1} - 1\right). \tag{7.56}$$

The ratio J_{ss}/J_C is plotted in Fig. 7.16(b) vs. γ_C/γ_0. If the material is brittle and fractures as soon as the yield point is reached, i.e. $\gamma_C/\gamma_0 = 1$, then $J_{ss}/J_C = 1$ and there is no resistance curve. In ductile materials such as most structural metals it would be expected that the strain to failure would be many times the yield strain and thus from Fig. 7.16(b) the steady state value of J would be much larger than the initiation value of J. Experimental data support this conclusion.

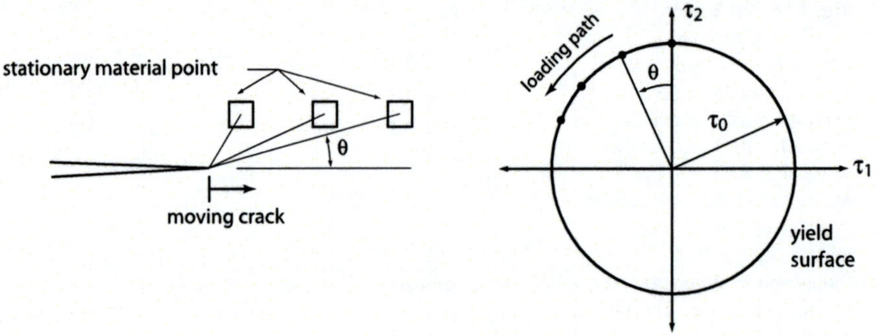

Fig. 7.17 (**a**) The angle θ of a stationary material point increases from $0 \to \pi/2$ as the crack passes the point. (**b**) The stress state stays on the yield surface, but moves from $\theta = 0 \to \theta = \pi/2$ as the crack grows. In a stationary crack the stress at a material point located at (x_1, x_2) with respect to the crack tip stays at the same point on the yield surface with $\theta = \tan^{-1} x_1/x_2$

The stress fields for the stationary and moving cracks are the same, namely $\tau_r = 0$ and $\tau_\theta = \tau_0$, so why are the strain fields different? The reason has to do with the non proportionality of the stresses in the moving crack case. In Cartesian coordinates the stresses are $\tau_1 = -\tau_0 \sin\theta$ and $\tau_2 = \tau_0 \cos\theta$. As the crack approaches a stationary material point θ varies from 0 to $\pi/2$. Thus the stress moves along the yield surface as shown in Fig. 7.17.

7.5.4 Transient Crack Growth in Elastic-Perfectly Plastic Material

The above shows that ductile materials can have J_{ss} values that are well above the J_C value needed for starting crack growth. How does J increase from J_C to J_{ss}? To answer this we consider the transient case of a crack just after it begins to grow. Rice [5] shows that for a transient crack the strain γ_θ ahead of the crack (on $\theta = 0$) is

$$\dot{\gamma}_\theta(r,a) = \gamma_0 \frac{\ln(r_p/r)}{r} + \frac{\gamma_0}{r}(1 + \dot{r}_p), \tag{7.57}$$

were $(\dot{\cdot})$ denotes differentiation with respect to crack length, a, and r is the distance from the crack tip. Based on the result for a stationary crack assume that the plastic zone size is related to J by

$$r_p = \frac{2}{\pi} \frac{J}{\tau_0 \gamma_0}. \tag{7.58}$$

As we did for steady-state crack growth we take the fracture criterion to be $\gamma_\theta = \gamma_c$ at $r = r_c$ on $\theta = 0$. Consider growth starting from a previously stationary crack. In this case, from Eq. (7.40) with $r_p = 2R$,

$$\frac{\gamma_\theta}{\gamma_0} = \frac{r_p}{r},$$

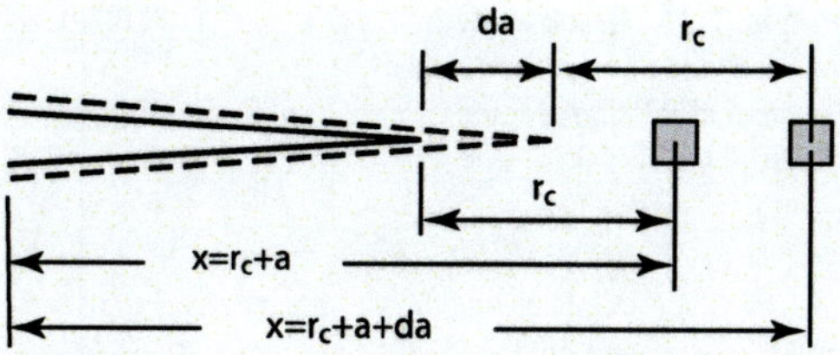

Fig. 7.18 Transient crack growth in anti-plane shear. As crack grows by length da, the strain at points $r = r_c$ ahead of the moving crack is γ_c

$$\frac{\gamma_c}{\gamma_0} = \frac{r_p}{r_c}.$$

Let x be the location of a material particle measured in a fixed coordinate system. The strain at a fixed material point is written as $\gamma_\theta = \gamma_\theta(x, a)$, where a is the crack length. Prior to crack growth $r = x$ and hence the strain can be written as

$$\gamma_\theta = (\gamma_c r_c)\frac{1}{x}. \tag{7.59}$$

Consider the strain at a distance r_c from the crack tip before and after an increment of crack growth, see Fig. 7.18.

$$\gamma_\theta(r_c + da, a + da) \approx \gamma_\theta(r_c, a) + \frac{\partial \gamma_\theta(r_c, a)}{\partial x}\frac{\partial x}{\partial a} da + \frac{\partial \gamma_\theta(r_c, a)}{\partial a} da.$$

Note that $\partial x/\partial a = 1$ and $\partial \gamma/\partial a = \dot{\gamma}$. Since the failure criterion must be satisfied $\gamma_\theta(r_c + da, a + da) = \gamma_c$ and $\gamma_\theta(r_c, a) = \gamma_c$. Thus, differentiating Eq. (7.59)

$$\frac{\partial \gamma_\theta}{\partial a} = \dot{\gamma}_\theta(r_c, a) = -\frac{\partial \gamma_\theta(r_c, a)}{\partial x} = \left.\frac{\gamma_c r_c}{x^2}\right|_{x=r_c} = \frac{\gamma_c}{r_c}. \tag{7.60}$$

Combining Eqs. (7.60) and (7.57) and evaluating the expression on $r = r_c$

$$\dot{\gamma}_\theta(r_c, a) = \frac{\gamma_c}{r_c} = \gamma_0 \frac{\ln(r_p/r_c)}{r_c} + \frac{\gamma_0}{r_c}(1 + \dot{r}_p).$$

Define $\alpha \equiv \gamma_c/\gamma_0$ and note that $\alpha = r_p/r_c$. Thus

$$\alpha = \ln\alpha + \dot{r}_p + 1,$$

or

$$\dot{r}_p = \alpha - \ln\alpha - 1.$$

Differentiating Eq. (7.58)

$$\dot{r}_p = \frac{2}{\pi}\frac{\dot{J}}{\tau_0\gamma_0}.$$

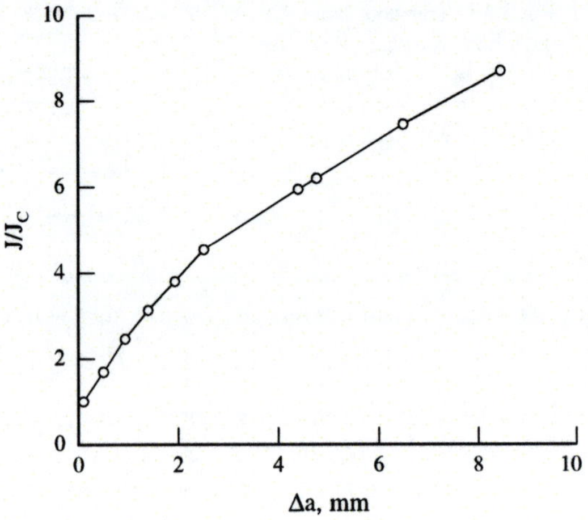

Fig. 7.19 Crack growth resistance curve for A533-B steel at 93°C. Adapted from [7]

Thus we the rate of change of J needed to grow the crack according to the critical strain criterion is

$$\dot{J} = \frac{\pi \tau_0 \gamma_0}{2}(\alpha - \ln \alpha - 1), \quad \text{or}$$
$$\frac{\dot{J}}{J_c} = \frac{1}{r_p}(\alpha - \ln \alpha - 1), \tag{7.61}$$

using Eq. (7.43) to relate the plastic zone and initiation value of J.

Since crack growth stability is related to the slope of the J resistance curve versus the slope of the available $J(a)$ it is seen that ductile materials have a steep resistance curve and will have extended tearing stability. For example consider $\alpha = 10$. From the steady state calculation $J_{ss}/J_c = 2.9$. The slope of the resistance curve is $\dot{J}/J_c = 6.7/r_p$. For higher ductility, $\alpha = 20$, the corresponding values are $J_{ss}/J_c = 9.5$ and $\dot{J}/J_c = 16/r_p$. Structural metals will generally have values of α that are in the range of 100, thus, this analysis suggests that such metals should have steeply rising J resistance curves. An example is shown in Fig. 7.19 where the J resistance curve is given for crack growth in A533-B steel tested at 93°C. This steep slope enhances the stability of crack propagation.

7.6 Mode-I Crack in Elastic-Plastic Materials

7.6.1 Stationary Crack in a Power Law Hardening Material

7.6.1.1 Deformation Theory (HRR Field)

We consider the field at the tip of a stationary crack in a power law hardening material. This solution is known as the HRR field after the authors Hutchinson, Rice and

Rosengren who published the result in 1968 [8, 9]. The crack is semi-infinite and the applied loads increase monotonically so that proportional loading is *approximated* allowing the use of the deformation theory of plasticity. The analysis assumes that the plastic strains are much greater than the elastic strains, thus $\gamma \sim \alpha\sigma^n$.

A great deal of information about the structure of the fields can be obtained without going into a detailed analysis. Since we are dealing with deformation theory of plasticity, which is actually a nonlinear elasticity theory, the J integral is path independent and $J \neq 0$ unless the crack is not loaded. Consider a circular path of radius r surrounding the crack tip. Shrinking this contour down to the crack tip

$$J = \lim_{r\to 0} \int_{-\pi}^{\pi} \left(Wn_1 - t_i \frac{\partial u_i}{\partial x_1}\right) r\,d\theta = \text{Const.}$$

Since the integrand is over θ only it must be constant with respect to r, thus

$$\lim_{r\to 0} \left(Wn_1 - t_i \frac{\partial u_i}{\partial x_1}\right) = \frac{f(\theta)}{r}.$$

Thus we expect

$$\lim_{r\to 0} \sigma \cdot \gamma \sim \frac{g(\theta)}{r}. \tag{7.62}$$

Expand σ in a power series about $r = 0$,

$$\sigma_{ij}(r,\theta) = \sum_s r^s \hat{\sigma}_{ij}(\theta, s).$$

Choosing the most singular term and noting that as $r \to 0$, $\|\gamma^p\| \gg \|\gamma^e\|$, from Eq. (7.32) the strain is

$$\gamma_{ij} = \gamma_{ij}^p = \frac{3}{2}\gamma_0\alpha\left(\frac{\sigma_e}{\sigma_0}\right)^{n-1}\frac{S_{ij}}{\sigma_0} = r^{sn}\hat{\gamma}_{ij}(\theta, s).$$

Knowing that $\sigma \cdot \gamma \sim 1/r$ we have $r^s \cdot r^{sn} \sim r^{-1}$, thus $s + sn = -1$, or

$$s = \frac{-1}{1+n}.$$

Thus as $r \to 0$ the stress and strain field may be written as

$$\begin{aligned} \sigma_{ij} &= \sigma_0 k_n r^{-1/(n+1)} \tilde{\sigma}_{ij}(\theta, n), \\ \gamma_{ij} &= \gamma_0 \alpha k_n^n r^{-n/(1+n)} \tilde{\gamma}_{ij}(\theta, n), \end{aligned} \tag{7.63}$$

where k_n is a plastic stress intensity factor, dependent on the hardening, n. In the linear elastic limit $n \to 1$ and the stress and strain as singular as $1/\sqrt{r}$. In the elastic perfectly plastic limit $r \to \infty$ and the strain is singular as $1/r$ while the stress is finite. Recall that in the stationary anti-plane shear crack the stress is finite while the strain is $1/r$ singular.

The angular dependence of the stress and strain fields are solved by introducing a stress function, ϕ, such that σ is given by derivatives of the function with respect to r and θ. Using the constitutive relation, Eq. (7.32) and compatibility (much like the solution of elastic crack problems) a differential equation for the stress function can be formed. Assuming separability of the stress function the r variation of σ and γ comes out as above, leaving a non-linear ordinary differential equation for the angular variation of ϕ. Applying the boundary conditions that $\sigma_{\theta\theta}(\pm\pi) = \sigma_{\theta r}(\pm\pi) = 0$ the equation is solved numerically using the shooting method. For large n values, the resulting angular variations agree closely with the elastic-perfectly plastic result, see Sect. 7.6.2 and Ref. [8] for the details.

The dependence of the stress and strain on the J integral value can also be determined. Substituting Eqs. (7.63) into the J integral and evaluating it on a circular contour

$$J = \int_{-\pi}^{\pi} \left(W n_1 + t_i \frac{\partial u_i}{\partial x_1} \right) r \, d\theta$$
$$= \sigma_0 \gamma_0 \alpha k_n^{n+1} \int_{-\pi}^{\pi} \tilde{f}_{ij}(\theta, n) \, d\theta,$$

where $\tilde{f}(\theta, n)$ contains the angular variations of the J integral terms as computed using the stress function analysis. The integral

$$I_n \equiv \int_{-\pi}^{\pi} \tilde{f}_{ij}(\theta, n) \, d\theta$$

is of order one and has been numerically tabulated. Thus the amplitude k_n can be written as

$$k_n = \left[\frac{J}{\alpha \sigma_0 \gamma_0 I_n} \right]^{\frac{1}{n+1}},$$

and

$$\begin{aligned} \sigma_{ij} &= \sigma_0 \left(\frac{J}{r \alpha \sigma_0 \gamma_0 I_n} \right)^{\frac{1}{n+1}} \tilde{\sigma}_{ij}(\theta, n), \\ \gamma_{ij} &= \gamma_0 \alpha \left(\frac{J}{r \alpha \sigma_0 \gamma_0 I_n} \right)^{\frac{n}{n+1}} \tilde{\gamma}_{ij}(\theta, n). \end{aligned} \tag{7.64}$$

Thus we see that near the tip of a stationary crack in an elastic-perfectly plastic material the magnitudes of the stress and strain field are governed by the value of the J integral.

J is the plastic analog of the stress intensity factor in LEFM. Assuming SSY conditions exist, $J = K_I^2/E'$ and hence the crack tip stress and strain magnitudes are determined by the stress intensity factor. This provides some insight into the application of K_I as a fracture criterion for materials that are ductile, but not so ductile that SSY does not apply. In such cases the stress intensity factor still governs the crack tip stress and strain singularities, although with different singularities than the elastic case, and thus is expected to be valid as a fracture initiation criterion. When SSY is not valid we may be able to use $J > J_c$ as a fracture initiation criterion as was discussed in the context of the anti-plane shear crack.

Note that the HRR analysis is only valid very near the crack tip, where the plastic strains are much larger than the elastic strains. This leads to the question of what is the region of dominance of the HRR fields, i.e. over how large of a region around the crack tip do the HRR fields provide a good description of the actual stress fields. Finite element results demonstrate that the region of dominance of the HRR fields decreases with increasing n, i.e. lower hardening, and vanishes when $n \to \infty$.

7.6.1.2 Incremental Theory

The above analysis of a crack in a material obeying deformation theory is an *approximation* to real world problems. Carka and Landis [10] have recently performed a series of detailed finite element studies of crack tip fields using an incremental, J_2 flow theory of plasticity with power law hardening. Their results show that in incremental theory J is *path dependent* and that the loading is non-proportional. Using the domain integral method, Sect. 5.4.4, the J integral was calculated over circular contours within the plastic zone, at varying distances from the crack tip. The results show that for a non-hardening material the J integral at the crack tip is approximately 18% less than the far-field J integral. As hardening increases (smaller values of hardening exponent n) the reduction in J drops, becoming less than 2% for $n < 1.4$. The degree of path dependence is also affected by the in-plane stress parallel to the crack line (T stress), by the out-of-plane stress level parallel to the crack front and by the mode mixity (proportion of Mode-I to Mode-II loads).

Path dependence may call into questions the use of the J integral as a crack tip loading parameter and fracture criterion. However as long as yielding is contained the far-field value of J still determines the deformation in the plastic zone and provides the crack tip driving force. When the plastic zone is no longer contained (see Sect. 7.6.3) there may not be a far-field value of J to refer to in computing the crack driving forces. In such cases other more physically based criteria such as crack tip opening displacement, or crack opening angle or a critical strain criterion may be more suitable.

7.6.2 Slip Line Solutions for Rigid Plastic Material

7.6.2.1 Introduction to Plane Strain Slip Line Theory

The slip line theory is useful for the construction of solutions of simple problems involving perfectly plastic materials. The slip line solutions for fracture problems will reveal the basic structure of the crack tip stress fields and provide limits on

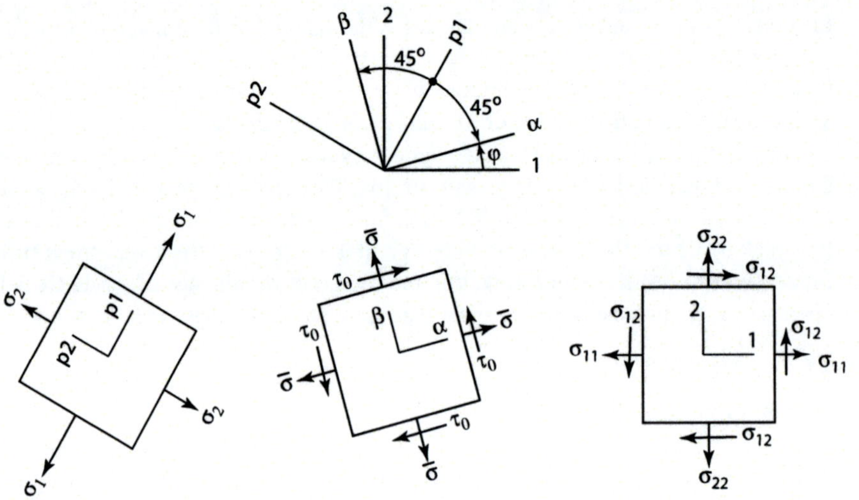

Fig. 7.20 Basis vectors and stresses in principal axes, in axes of maximum shear stress and in $\mathbf{e}_1, \mathbf{e}_2$ axes

the crack tip normal (σ_{22}) and hydrostatic (σ_{kk}) stresses. These stresses will be a multiple of the flow stress. Higher crack tip stresses occur when the material near the crack tip is highly "constrained" such as in plane strain where no contraction in the direction parallel to the crack line is allowed. We will discuss only the plane strain theory as it is simpler than the plane stress theory.

In plane strain slip line theory is assumed that γ_{11}, γ_{22} and γ_{12} are the only non-zero strains. The stresses $\sigma_{\alpha\beta}$ and σ_{33} are functions of x_1, x_2 and $\sigma_{\alpha 3} = 0$. The in-plane displacement rates $v_\alpha = v_\alpha(x_1, x_2)$ and $v_3 = 0$ due to the plane strain constraint. It is assumed that the plastic strains are much larger than the elastic strains.

Using the von-Mises yield condition, the plane strain constraint results in $\sigma_{33} = \frac{1}{2}(\sigma_{11} + \sigma_{22})$. The yield condition can be written as

$$\left(\frac{\sigma_{11} - \sigma_{22}}{2}\right)^2 + \sigma_{12}^2 = \tau_0^2, \tag{7.65}$$

where τ_0 is the flow stress in pure shear. In terms of the Mohr's circle the yield condition be represented as a circle of radius τ_0 centered on the mean stress, $\bar{\sigma} = (\sigma_{11} + \sigma_{22})/2$.

Let $(\mathbf{p}_1, \mathbf{p}_2)$ be the principal directions, see Fig. 7.20. The planes of maximum shear stress occur at $\pm 45°$ from $\mathbf{p}_1, \mathbf{p}_2$. Let the $\mathbf{e}_\alpha$ be the direction at $-45°$ from $\mathbf{p}_1$, and let $\mathbf{e}_\beta$ be the direction at $+45°$ from $\mathbf{p}_1$. In the $(\mathbf{e}_\alpha, \mathbf{e}_\beta)$ basis the normal stresses are $\bar{\sigma}$ and the shear stress is τ_0.

Using the Mohr's circle, the stress components in the $(\mathbf{e}_1, \mathbf{e}_2)$ basis can be written in terms of $\bar{\sigma}$, τ_0 and the rotation angle, φ, needed to transform from (x_1, x_2) axes to the axes of maximum shear stress,

$$\begin{aligned}\sigma_{11} &= \bar{\sigma} - \tau_0 \sin 2\varphi,\\ \sigma_{22} &= \bar{\sigma} + \tau_0 \sin 2\varphi,\\ \sigma_{12} &= \tau_0 \cos 2\varphi.\end{aligned} \tag{7.66}$$

The stress field is thus defined by the $\varphi(x_1, x_2)$ and $\bar{\sigma}(x_1, x_2)$ fields. The direction of the $\mathbf{e}_1, \mathbf{e}_2$ basis vectors changes from point to point in the field, but one can consider a network of orthogonal α and β lines that give the directions of $\mathbf{e}_\alpha, \mathbf{e}_\beta$. Along these lines it can be shown that

$$\begin{aligned}\bar{\sigma} - 2\tau_0\varphi &= \text{constant on } \alpha \text{ lines},\\ \bar{\sigma} + 2\tau_0\varphi &= \text{constant on } \beta \text{ lines}.\end{aligned}$$

Relations between the velocity fields can be derived as well, but we will concern ourselves here only with the stress fields.

To understand how the slip line fields work consider the following very simple example. Assume that a block of material is loaded in compression with $\sigma_{22} = -p$, $\sigma_{11} = \sigma_{12} = 0$ as shown in Fig. 7.21(a). From $\sigma_{12} = 0$, we know that $\varphi = \pm 45°$. The (x_1, x_2) directions are also the principal stress directions. Knowing that the α direction is found by a $-45°$ rotation from the direction of maximum principal stress, x_1 direction in this case, we know that $\mathbf{e}_\alpha$ is as shown in Fig. 7.21(a) and thus $\varphi = -45°$. Then $\sigma_{11} = 0 = \bar{\sigma} - \tau_0 \sin(-90°) = \bar{\sigma} + \tau_0$, thus $\bar{\sigma} = -\tau_0$. Then $\sigma_{22} = -p = \bar{\sigma} + \tau \sin(-90°) = -\tau_0 - \tau_0$, or $p = 2\tau_0$. The α and β lines form a network at $\pm 45°$.

The geometry will not always be so simple, so a few rules are in order. (1) Consider the α and β lines shown in Fig. 7.21(b). It can be shown that

$$\varphi_A - \varphi_B = \varphi_D - \varphi_C. \tag{7.67}$$

(2) The change in φ along any β lines between two particular α lines is constant. The change in φ along any α lines between two particular β lines is constant. (3) If one β line is straight between two α lines, then all β lines are straight between these two α lines. (4) If the α and β lines are both straight, then the stress is constant (as in the example). (5) The stress is constant along a straight β or α line.

7.6.2.2 Plane-Strain, Semi-infinite Crack

Consider a semi-infinite crack under tensile loading. Boundary conditions on the crack faces are $\sigma_{22} = \sigma_{12} = 0$. The stress field (Prandtl solution) has three sectors, consisting of constant stress regions ahead of (region III) and behind the crack (region I) and a centered fan (region II), see Fig. 7.22. In region I if $\varphi = -\pi/4$, then the boundary conditions are satisfied if $\bar{\sigma} = \tau_0$ resulting in $\sigma_{11} = 2\tau_0$, $\sigma_{22} = 0$, $\sigma_{12} = 0$ in *region I*.

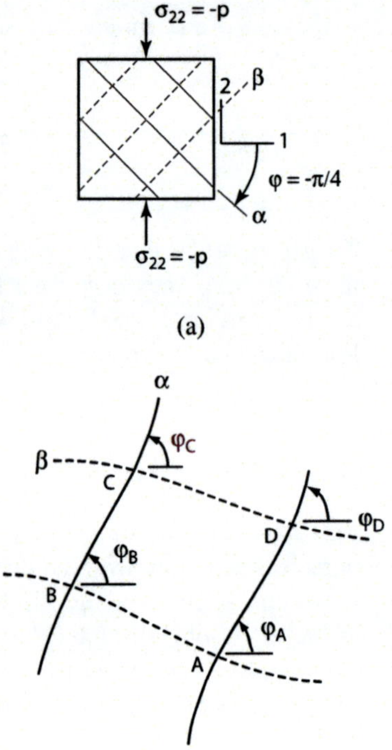

Fig. 7.21 (**a**) Slip line field in the case of $\sigma_{22} = -p$. (**b**) General network of α and β slip lines

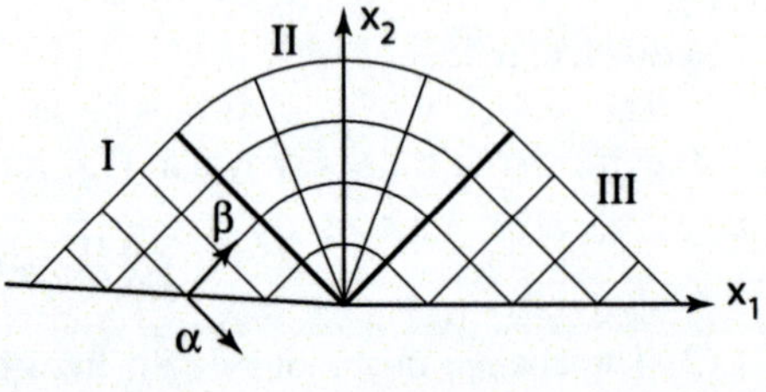

Fig. 7.22 Prandtl slip line field for plane strain crack in rigid plastic material. Stress solution from HRR as $n \to 0$ field agrees with slip line field

Follow the β line from region I into region III. Along this line $\bar{\sigma} + 2\tau_0\varphi = \text{const.}$ In region I this value is $\bar{\sigma} + 2\tau_0\varphi = \tau_0(1 - \pi/2)$. In region III $\varphi = -3\pi/4$, thus $\bar{\sigma} + 2\tau_0\varphi = \tau_0(1 - \pi/2) = \bar{\sigma} + 2\tau_0(-3\pi/4)$, or $\bar{\sigma} = \tau_0(1 + \pi)$. Applying Eqs. (7.66) the stresses are then $\sigma_{22} = \tau_0(2 + \pi)$, $\sigma_{11} = \tau_o\pi$, $\sigma_{12} = 0$ in *region III*.

In the centered fan, region II, $\varphi = \theta - \pi$ where θ is the angle of a material point with respect to the x_1 axis. Using the result that $\bar{\sigma} + 2\tau_0\varphi = \text{constant} = \tau_0(1 - \pi/2)$ along a β line, $\bar{\sigma} = \tau_0(3\pi/2 + 1 - 2\theta)$. Applying Eqs. (7.66) the stresses are $\sigma_{22} = \tau_0(3\pi/2 + 1 - 2\theta + \sin 2\theta)$, $\sigma_{11} = \tau_0(3\pi/2 + 1 - 2\theta - \sin 2\theta)$, $\sigma_{12} = \tau_0 \cos 2\theta$ in *region II*. The stresses are plotted in Fig. 7.23.

Let us probe these results. Under the von-Mises yield criterion the yield stress in shear, τ_0 and the yield stress in tension, σ_0 are related by $\tau_0 = \sigma_0/\sqrt{3} = 0.577\sigma_0$. In terms of σ_0 the normal stress ahead of the crack is $\sigma_{22} = 2.97\sigma_0$. The hydrostatic stress is $\sigma_{kk} = 3\bar{\sigma} = 7.17\sigma_0$. The T-stress, $\sigma_{11} = 1.18\sigma_0$. Thus in plane strain

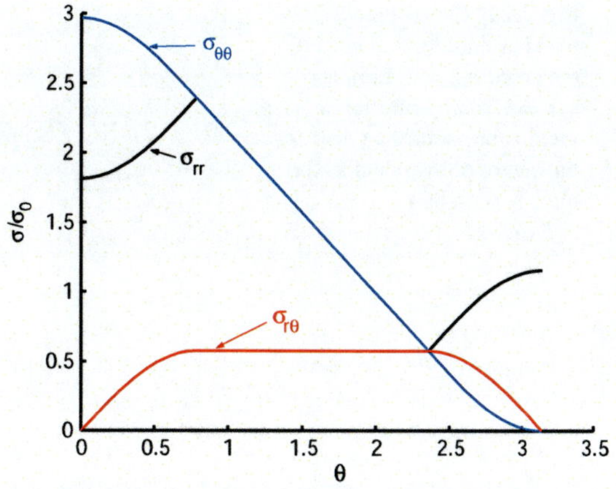

Fig. 7.23 Angular distribution of stresses for the Prandtl slip line field ahead of a tensile crack in a perfectly plastic material

the tensile and hydrostatic stresses are considerably elevated above the uniaxial tensile values due to the "constraint" imposed by the plane strain ($\gamma_{33} = 0$) and by the constraint imposed by the surrounding material that does not allow significant contraction of material in the x_1 direction ahead of the crack tip.

7.6.2.3 Plane-Stress, Semi-infinite Crack

The analysis of slip line fields in plane stress is somewhat more complex, thus only the results will be quoted here. In plane stress, ahead of the crack tip $\sigma_{22} = 2\tau_0 = 1.15\sigma_0$, $\sigma_{11} = \tau_0 = 0.57\tau_0$ and $\sigma_{kk} = 3\tau_0 = 1.7\sigma_0$. Thus in comparison to plane strain, the relative lack of constraint, i.e. γ_{33} is not restricted, hence $\sigma_{33} = 0$, results in much lower stress values at the crack tip.

7.6.3 Large Scale Yielding (LSY) Example

If SSY conditions are no longer maintained and the plastic zone stretches across a significant fraction of the remaining ligament then "large scale yielding" or LSY conditions prevail. In LSY the constraint can change considerably, modifying the crack tip fields relative to the SSY result above rendering the concept of autonomy to be inapplicable. Consider the center crack panel shown in Fig. 7.24 loaded to its limit load, i.e. the plastic zone stretches across the entire uncracked ligament and the applied load can no longer be increased. A slip line field can be constructed consisting of constant stress regions with $\bar{\sigma} = \tau_0$ and $\varphi = \pi/4$, resulting in $\sigma_{22} = 2\tau_0 = 1.15\sigma_0$ and $\sigma_{kk} = 3\tau_0 = 1.73\sigma_0$ ahead of the crack tip. In the plastic regions the material is free to strain by moving towards the center of the plate. Strain concentrates along the boundary of the yield and elastic zones. The reduction in constraint results in a large reduction in the tensile and hydrostatic stress ahead of the crack.

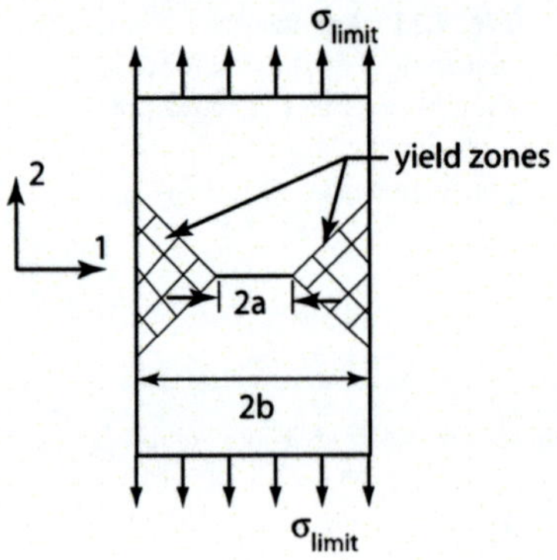

Fig. 7.24 Center cracked panel at limit load. Ahead of the crack $\sigma_{22} = 1.15\sigma_0$ and $\sigma_{kk} = 1.73\sigma_0$. Material in yield zones deforms towards the center of the panel as the plate is stretched

As the above simple example shows the stress and strain under LSY conditions can be much different from the SSY fields and can be strongly configuration (i.e. center cracked plate, bar in bending [11], edge cracked plate) dependent. Under such conditions autonomy is not expected to be valid and fracture toughness values, e.g. J_C, measured in a laboratory test configuration may not be valid for a LSY application to a different configuration.

7.6.4 SSY Plastic Zone Size and Shape

Fundamental to the understanding of elastic-plastic fracture and to the analysis of the limitations of the SSY approach is the determination of the size and shape of the crack tip plastic zone. We have already seen that in LSY the size may extend across the entire uncracked ligament. Furthermore the shape and distribution of plastic strain within the plastic zone can be highly configuration dependent. Let us concentrate on 2D cracks under SSY conditions in plane stress and plane strain.

Define r_p as the extent of the plastic zone directly ahead of the crack, i.e. along $\theta = 0$. We know from the Dugdale model and from the Mode-III solution that r_p will scale as $r_p = c(K_I/\sigma_0)^2$. However the value of c must be determined for different levels of hardening for plane stress and plane strain. It does not suffice to simply take the Williams K_I field and substitute it into the yield condition, doing so will underestimate the plastic zone, however one can approximate the shape of the plastic zone by such an analysis. Using the von-Mises condition estimates for the plastic zone shapes for SSY for $\nu = 0.25$ are shown in Fig. 7.25.

Define r_p to be the extent of the plastic zone directly ahead of the crack tip. Assume that for a non-hardening material the stress ahead of the crack is as shown in Fig. 7.26 consisting of a yielded region with $\sigma_{22} = \eta\sigma_0$ of length r_p. The elastic field outside the plastic zone is a K_I field shifted forward by $r_p - r_y$, where r_y is the distance ahead of the shifted crack at which the shifted elastic K_I field and the plastic fields cross, i.e.

$$\eta\sigma_0 = \frac{K_I}{\sqrt{2\pi r_y}},$$

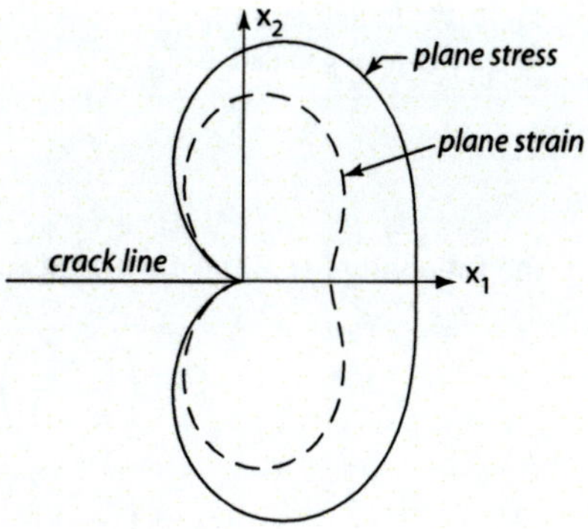

Fig. 7.25 Shapes of the plane stress and plane strain plastic zones estimated by plugging the K_I field into the von-Mises yield criterion. Actual SSY plastic zone shapes and sizes will depend on n, T, and ν

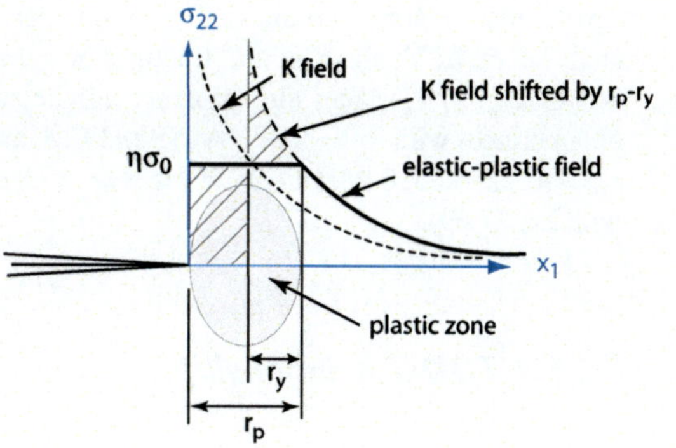

Fig. 7.26 Schematic of stress ahead of crack in elastic-plastic non-hardening material

or

$$r_y = \frac{1}{2\pi}\left(\frac{K_I}{\sigma_0}\right)^2.$$

The load carried by the shifted K_I field in the shaded region above the plastic zone, from $r_p - r_y$ to r_y, must be carried in the plastic zone from 0 to $r_p - r_y$. Thus

$$\eta\sigma_0(r_p - r_y) = \int_0^{r_y} \left(\frac{K_I}{\sqrt{2\pi r}} - \eta\sigma_0\right) dr,$$

resulting in

$$r_p = 2r_y = \frac{1}{\pi}\left(\frac{K_I}{\eta\sigma_0}\right)^2. \tag{7.68}$$

For plane stress the slip line field suggests $\eta = 1.2$. However considering the effects of crack tip blunting the value may be closer to $\eta = 1$. In plane strain the slip line field suggests $\eta = 2.97$, however in consideration of crack tip blunting (see Sect. 7.6.8) and three dimensional effects the value $\eta = \sqrt{3}$ is often taken for plane strain. Using these values of η and making a correction for hardening [12] we have

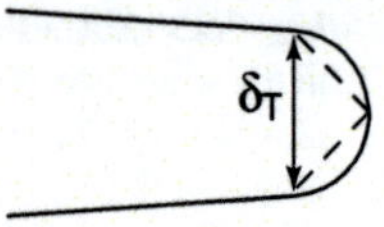

Fig. 7.27 Definition of δ_T for elastic-plastic materials

the following *estimates* for the plastic zone size:

$$r_p = \left(\frac{n-1}{n+1}\right)\frac{1}{\pi}\left(\frac{K_I}{\sigma_0}\right)^2 \quad \text{plane stress,} \tag{7.69}$$

$$r_p = \left(\frac{n-1}{n+1}\right)\frac{1}{3\pi}\left(\frac{K_I}{\sigma_0}\right)^2 \quad \text{plane strain.} \tag{7.70}$$

Note that a small strain (i.e. no consideration of crack tip blunting) FEM analysis of SSY plane strain fracture in a non-hardening material yields $r_p \approx 0.036(K_I/\sigma_0)^2$ [13] an almost exact agreement with the r_p estimate that would be obtained with $\eta = 2.97$. A SSY FEM analysis of plane stress [14] yields $r_p = (0.22, 0.25, 0.28)(K_I/\sigma_0)^2$ for $n = (5, 9, \infty)$ respectively, in close agreement with Eq. (7.69).

7.6.5 CTOD-J Relationship

The crack tip opening displacement, δ_T can be measured experimentally and the attainment of a critical value of δ_T may under certain conditions be useful as a fracture criterion. There are a number of definitions of δ_T. Let us define δ_T as the opening across the crack where ±45° lines from the crack tip intercept the crack surfaces, see Fig. 7.27.

In the Dugdale model we found that the crack tip opening displacement δ_T was related to J by $\delta_T = J/\sigma_0$. A similar expression was found for the anti-plane shear crack. Analyses of the HRR field and finite element studies exploring the δ_T-J relationship show that in SSY [15]

$$\delta_T = d_n \frac{J}{\sigma_0}, \tag{7.71}$$

where d_n is a constant dependent on n and on whether we have plane strain or plane stress conditions and σ_0 is the initial yield stress. The constant is also weakly dependent on the relative strength of the material, expressed as σ_0/E. Values of d_n are tabulated in Table 7.1. Computed crack opening profiles are plotted in Fig. 7.28 for $n = 1, 3, 13, \infty$. As hardening reduces the crack opening profile assumes a blunted shape.

In LSY the relation between δ_T and J is dependent on the configuration and on the extent of yielding. For example for $n = 10$, for a fully yielded, cracked bend specimen (see figure in Table 5.1) $d_n \approx 0.45$–0.40 and decreases as the loading

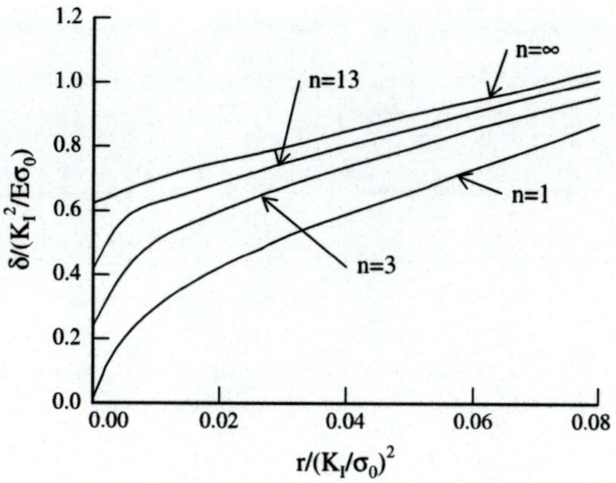

Fig. 7.28 Plane strain crack opening computed from FEM analysis of SSY problem. Adapted from [15]. Similar results are obtained for plane stress [14]

Table 7.1 Values of d_n computed using FEM with $\sigma_0/E = 0.002$ [15]

n	d_n, pl. strain	d_n, pl. stress
3	0.20	0.19
5	0.32	0.38
10	0.52	0.67
25	0.65	0.86
∞	0.70	1.0

continues. For a center cracked plate with $n = 10$, $d_n \approx 0.45$–0.60 and increases as loading continues. This is another example in which the LSY fields differ from the corresponding SSY values.

7.6.6 Growing Mode-I Crack

Depending on the geometry and loading, slow, stable crack growth is often observed prior to fracture instability. The analysis of crack growth up to the point of instability is critical to the calculation of the maximum load carrying capacity of a structure or component that is fractured. Thus the understanding of crack tearing is important in engineering practice. As with the moving Mode-III crack, the stresses for moving Mode-I cracks are quite similar to the fields for stationary cracks. However, due to non-proportional loading the strains for moving cracks have weaker singularities than their stationary counterparts.

For a plane strain crack in an elastic-perfectly plastic material the plastic strain ahead of the crack is given by [18]

$$\gamma^p_{\alpha\beta} = \frac{(5-4\nu)}{2\sqrt{2}}\frac{\tau_0}{E} G_{\alpha\beta}(\theta)\ln\left(\frac{R}{r}\right) + H_{\alpha\beta}(\theta), \quad r \to 0, \tag{7.72}$$

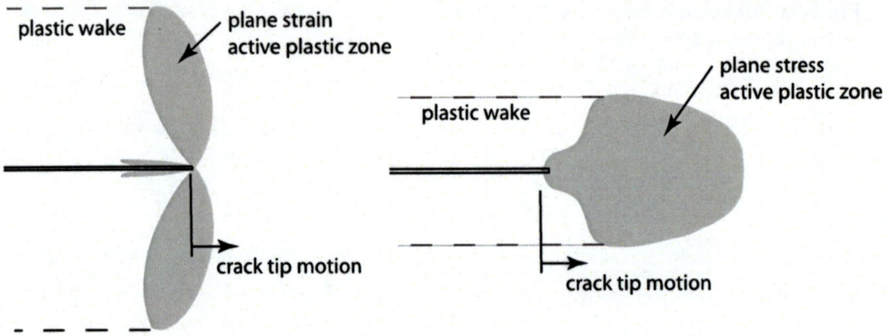

Fig. 7.29 Schematic of plastic zones for steady-state slow crack growth in elastic perfectly plastic materials. Actual plastic zones will depend on hardening, T stress and constraint. The active plastic zone is the region in which $\dot{\gamma}^p \neq 0$. As the crack advances it leaves behind a wake of residual deformation. This wake is often visible. Plane strain adapted from [16]. Plane stress adapted from [17]

where R is a length scale that is expected to scale with the plastic zone size, $H_{\alpha\beta}$ is undetermined by asymptotic analysis and $G_{\alpha\beta}$ is

$$G_{11}(\theta) = -G_{22}(\theta) = -2\sin\theta,$$

$$G_{12}(\theta) = \ln\left(\frac{\tan(\theta/2)}{\tan(\pi/8)}\right) + 2\left(\cos\theta - \frac{1}{\sqrt{2}}\right).$$

As with the moving Mode-III crack the strain is $\ln(1/r)$ singular, weaker than the corresponding stationary crack case in which $\gamma \sim 1/r$.

For a plane stress crack in an elastic perfectly plastic material the strain on $\theta = 0$ is [19]

$$\gamma_{ij}^p = \frac{1}{2}\frac{\tau_0}{E} G_{ij} \ln^2\left(\frac{R}{r}\right), \quad r \to 0, \tag{7.73}$$

where $G_{11} = 0$, $G_{12} = 0$, $G_{22} = 2$, $G_{33} = -2$. The strains in plane stress have a somewhat stronger $\ln^2(1/r)$ singularity than the plane strain case.

As the crack grows the plastic zone moves with the crack tip, sweeping through the material to leave a plastic wake, see Fig. 7.29. The wake will contain locked in strains that can often be seen by eye, at least on ductile metallic samples. In both plane strain and plane stress the crack opening profile during crack growth has a nearly constant angle as sketched in Fig. 7.30. The essentially constant crack opening angle forms that basis of the CTOA criterion that has been successfully employed for the prediction of crack tearing in sheets, see Sect. 8.4.3.

Assuming for fracture initiation that a critical crack opening displacement $\delta_T = \delta_C$ must be reached, the critical value of J for crack initiation is

$$J_C = \delta_C \sigma_0 / d_n.$$

For the moving crack the crack opening displacement is zero at the crack tip. Thus the failure criteria must be expressed in another manner. One approach is to assume

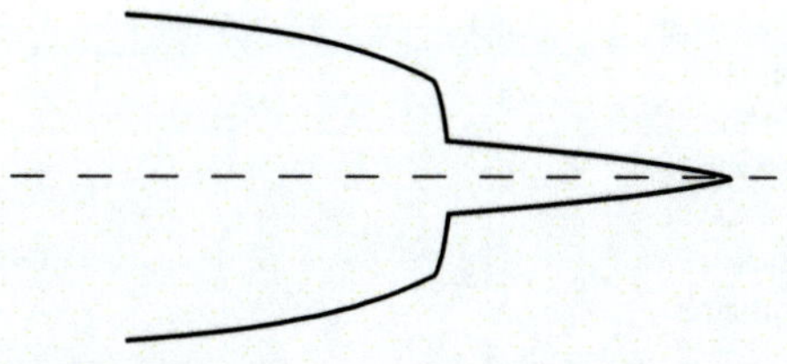

Fig. 7.30 Crack opening profile for a plane stress crack in an elastic-perfectly plastic material that grows from a blunted, stationary crack. Once fracture initiates the crack tip is relatively sharp and the crack grows with an approximately constant crack opening angle. The effect of hardening is to round off the crack opening somewhat. Vertical scale is greatly exaggerated. Adapted from [17]

that the crack grows when the crack opening angle (CTOA) reaches a critical value, i.e. the crack grows when the crack opening reaches a critical value, δ_C at a critical distance behind the crack tip, r_C,

$$\tan(\text{CTOA}) = \frac{\delta_C}{r_C} = \lambda_m \gamma_0.$$

The factor λ_m determines the ductility of the material.

Applying the above crack growth condition, Rice [18] calculates the J resistance curve, i.e. the value of J as a function of crack growth, Δa, from

$$\frac{dJ}{da} = \frac{\sigma_0^2}{E}\left(T_0 - \frac{\beta}{\alpha_{ssy}}\ln\left(\frac{J}{J_C}\right)\right),$$

where T_0 is the "tearing modulus", the initial normalized slope of the J vs. Δa curve,

$$T_0 \equiv \frac{E}{\sigma_0^2}\frac{dJ}{da}\bigg|_{\Delta a=0},$$

$\alpha_{ssy} \approx 0.65$, and $\beta = 5.083$. The tearing modulus can be related to δ_C/r_C by

$$T_0 = \frac{E}{\alpha_{ssy}\sigma_0}\frac{\delta_C}{r_C} - \frac{\beta}{\alpha_{ssy}}\ln\left(\frac{e\lambda E J_C}{r_C\sigma_0^2}\right),$$

where $\lambda = R/(EJ_C/\sigma_0^2) \approx 0.2$ relates the plastic zone size at crack initiation to J at crack initiation. The above can be integrated for various values of T_0 with the initial value $J(0) = J_C$. The results are plotted as J/J_C versus $\Delta a/R$ in Fig. 7.31. The plot shows that for ductile materials, (large T_0) the value of J needed to continue to drive the crack increases rapidly, see Figs. 8.9(a) and 8.9(b) for experimental data that show steep J resistance curves. Note that for brittle materials, $T_0 = 0$, $J = J_C$, i.e. the resistance curve will be flat.

Narasimhan et al. [17] show that in the elastic-perfectly plastic, plane stress, case the steady state value of J needed to propagate the crack is

$$\frac{J_{ss}}{J_C} = \frac{d_n}{s\lambda_m} e^{(\lambda_m/\beta - 1)}, \tag{7.74}$$

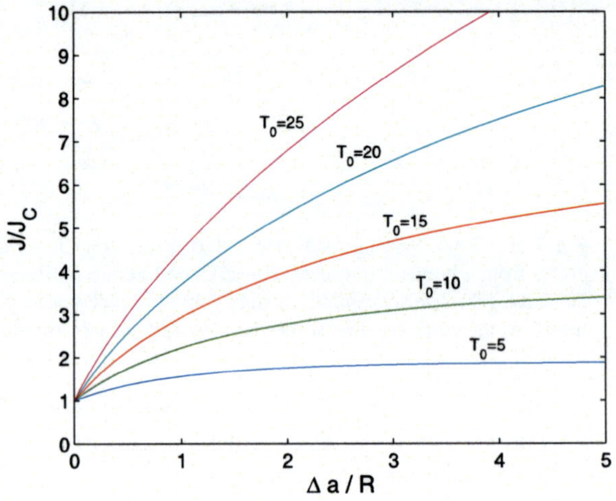

Fig. 7.31 *J* resistance curve predicted for plane strain using the criterion that the crack grows with constant crack opening angle. *R* is the plastic zone size at crack initiation. J_C is the initiation value of *J*

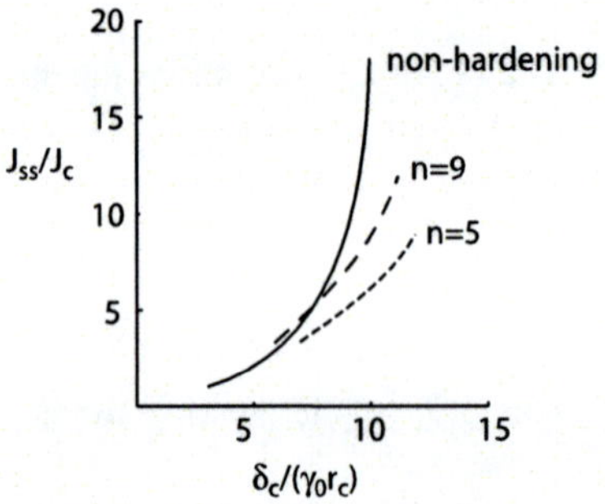

Fig. 7.32 J_{ss}/J_C for crack growth in elastic-plastic material under plane stress. Adapted from [17]

where $\beta = 1.70$ and $s = 0.60$ are parameters determined through a finite element analysis. The resulting J_{ss}/J_C is plotted in Fig. 7.32. Note that as with the Mode-III crack, the value of J needed to drive the crack increases rapidly with increasing material ductility. Also shown on the figure are J_{ss} values computed from finite element analyses of power law hardening materials. Note that for the same λ_m values the hardening materials have a lower toughness (J_{ss}), suggesting that lower stresses and greater ease of plastic flow in the non-hardening materials enhance fracture resistance, at least from the SSY point of view.

As a final word, note that in the real 3D world, ductile cracks do not grow on the flat plane $x_2 = 0$. Rather they form slant fractures in which the fracture surface slants at 45° to the plate, or V-shapes 45° to the plate surface, or surfaces that are flat in the center with 45° shear zones on the free surfaces. Although the actual fracture surfaces are far more complex that those assumed in the above analyses, these results do provide valuable insight into the origins of stable crack growth in ductile materials and provide hints on criteria such as CTOA that can be used to analyze ductile crack growth.

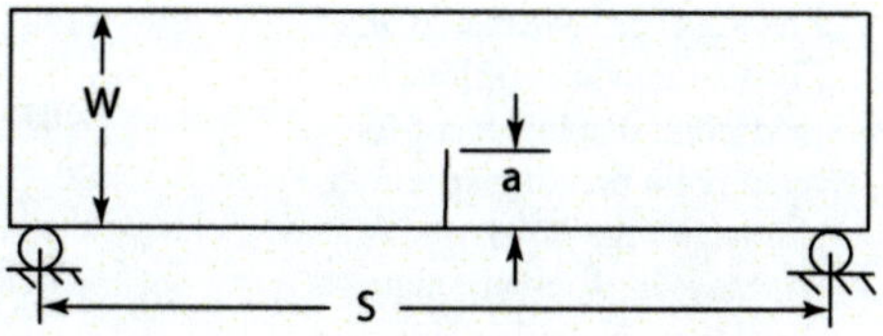

Fig. 7.33 Three point bend test specimen geometry. Dimensions: (see Fig. 5.1 for definitions) $B = 10$ mm, $W = 76$ mm, $a = 30$ mm, $S = 305$ mm. Adapted from [20]

7.6.7 Three Dimensional Aspects

Due to the strong differences between the plane stress and plane strain stress fields, the crack tip fields and plastic zones are strongly influenced by the three dimensional nature of the fields in test specimens and applications. As one example, consider the three point bending specimen shown in Fig. 7.33. The material is an annealed 4340 steel with $n = 22.5$, $\sigma_0 = 1030$ MPa [20, 21]. Experiments measuring load vs. load-point displacement and measuring the out-of-plane deflection of the surface of the sample were carried out in conjunction with a 2D/3D finite element study that analyzed the test using plane stress, plane strain and 3D FEM.

Experimental values of the J integral were calculated from [22]

$$J = \frac{2}{B(W-a)} \int_0^{\delta_c} P \, dv_{Lc}, \tag{7.75}$$

where P is the load, B is the thickness, $W - a$ is the uncracked ligament length, and v_{Lc} is the load-point displacement "due to the crack", i.e., $v_{Lc} = v_L - v_{Lnc}$, where v_L is the measured displacement and v_{Lnc} is the displacement of an uncracked, elastic beam of the same dimensions as the fracture specimen. Comparison of the experimental J values with the FEM results, see Fig. 7.34(b) shows that a 3D analysis is needed to accurately calculated J, but that the plane stress result is more accurate than the plane strain result since the plate is thin compared to its in-plane dimensions.

Through the thickness values of $J^*(x_3)$, are shown in Fig. 7.35 for different values of P/P_0, where P_0 is the plane stress limit load, i.e. the maximum load that can be obtained if the material were perfectly plastic [11]. As the loading continues and the sample transitions from SSY to LSY the crack tip constraint is relaxed somewhat, thus $\hat{J}$ drops off more quickly than the elastic case as the free surfaces are approached.

The degree of plane strain constraint can be judged by the value of $\sigma_{33}/(\sigma_{11} + \sigma_{22})$. For plane strain, $\sigma_{33}/\sigma_{\alpha\alpha} = 0.5$. For plane stress $\sigma_{33}/\sigma_{\alpha\alpha} = 0$. The through the thickness values of the plane constraint are plotted in Fig. 7.36 for $P/P_0 = 0.7$ as a function of distance from the crack tip, r/B and through the thickness location, x_3/B. Near the free surface, $x_3/B = 0.475$, $\sigma_{33}/\sigma_{\alpha\alpha} = 0$ as expected. Near the mid-plane, $x_3/B = 0.075$, $\sigma_{33}/\sigma_{\alpha\alpha} = 0.5$ very near the crack tip. However the constraint drops to zero for $r/B > 0.5$, thus the stress field at this distance from the crack tip can be described as a plane stress field.

A similar analysis was performed for crack in an elastic-plastic power-law hardening material [23] in which a circular region in a plate of thickness B (radius of

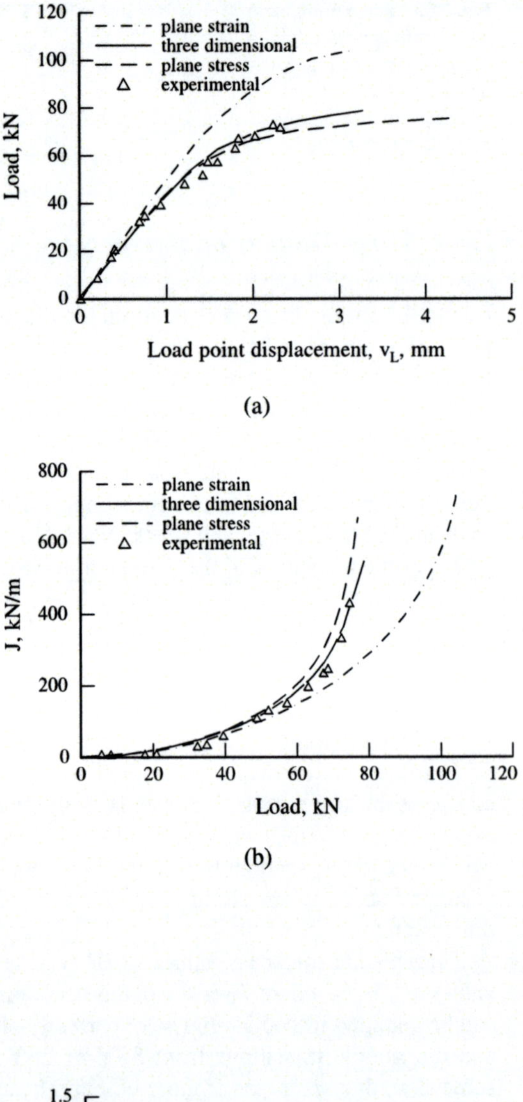

Fig. 7.34 (**a**) Measured and FEM simulations of load vs. load-point displacement for three point bending specimen shown in Fig. 7.33.
(**b**) Measured and simulated J vs. load. Note that since the test specimen is relatively thin the plane stress results more closely match the experimental results, but that the 3D simulation provides the best match

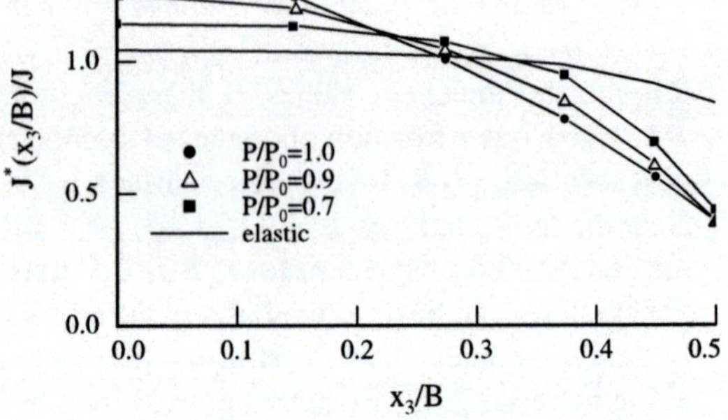

Fig. 7.35 Variation of J^* through the thickness for different load levels. P_0 is the plane stress limit load. From [20]

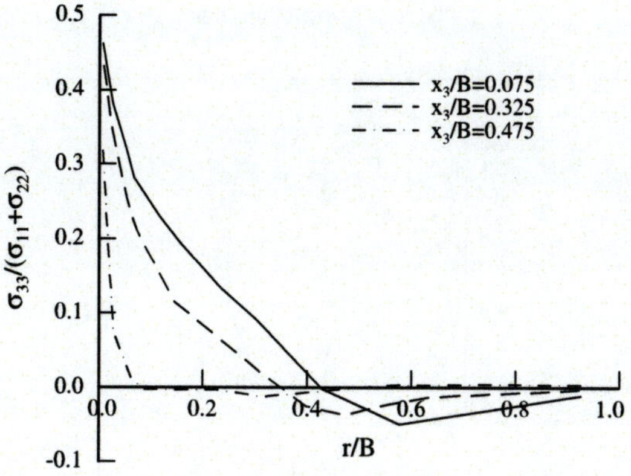

Fig. 7.36 $\sigma_{33}/(\sigma_{11}+\sigma_{22})$ ahead of the crack for the three point bend specimen shown in Fig. 7.33 at a load of $P/P_0 = 0.7$, where P_0 is the perfectly plastic limit load. The results show that very close to the crack tip, near the free surface ($z/B = 0.475$) σ_{33} quickly drops to zero. At the specimen mid-plane, σ_{33} drops to zero for $r/B > 0.5$. Thus beyond this distance the 3D problem is well approximated as a plane stress problem. Adapted from [20]

region is much larger than B) is analyzed using the plane stress K_I displacement field, Eq. (2.58) as the boundary condition. Plastic zones from [23] and [20] are shown and discussed in Fig. 7.37.

7.6.8 Effect of Finite Crack Tip Deformation on Stress Field

As an elastic-plastic material containing a Mode-I crack is loaded the crack tip will become blunted as shown in Fig. 7.27. The traction must be zero along the surface of this blunted crack, thus considering the geometry of the blunted crack, the traction free boundary condition implies that $\sigma_{11} = 0$ ahead of the crack. However, note that in the slip line field and in the HRR field, σ_{11} $(r, \theta = 0) \neq 0$. For the slip line field $\sigma_{11} = \tau_0 \pi = 1.8\sigma_0$. Forcing $\sigma_{11} = 0$ ahead of the crack will reduce the immediate crack tip stress levels. Using finite deformation theory, McMeeking [24] showed that for both hardening and non-hardening materials the stress is reduced below the HRR prediction in a region ahead of the crack that extends for approximately $2 - 3\delta_T$. The peak tensile stress ahead of the crack occurs at approximately $3\delta_T$ ahead of the crack, see Fig. 7.38. In the region $r < 3\delta_T$ the J integral is no longer path independent. It is shown that $J \to 0$ as the contour is shrunk to the blunted crack tip. For $r > 3\delta_T$, J is path independent and agrees with the value computed based on the applied loading. One implication of this result is that if fracture depends on the attainment of critical stresses these will be reached at a finite distance ahead of the crack.

The combined effects of finite deformation and three dimensionality were explored in a finite element study [25]. Figure 7.39 shows a cross section though the crack plane for an elastic plastic sheet with $n = 10$. Note that the crack front has

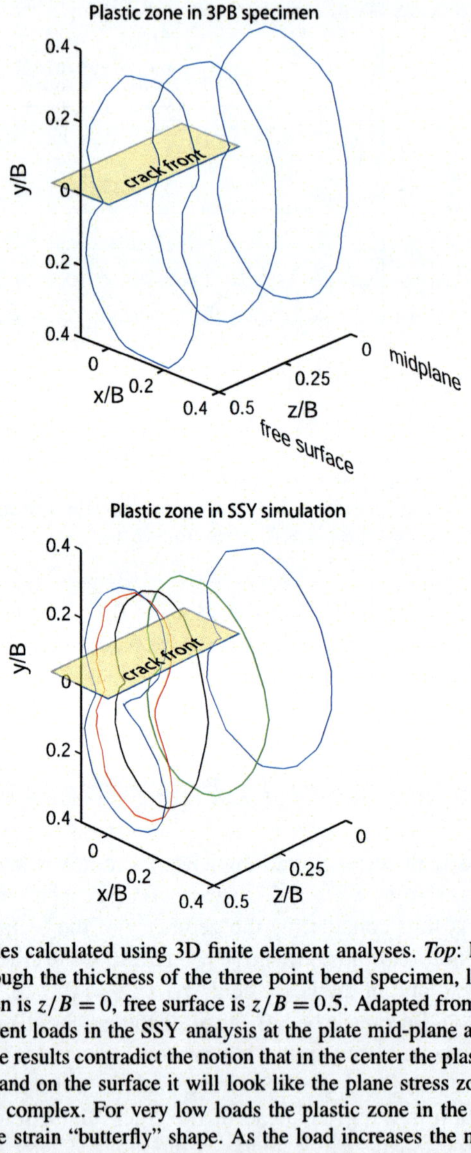

Fig. 7.37 Plastic zones calculated using 3D finite element analyses. *Top*: Plastic zone boundary at three locations through the thickness of the three point bend specimen, loaded at $P/P_0 = 0.7$. Mid-plane of specimen is $z/B = 0$, free surface is $z/B = 0.5$. Adapted from [20]. *Bottom*: Plastic zone shapes for different loads in the SSY analysis at the plate mid-plane and at the free surface, adapted from [23]. The results contradict the notion that in the center the plastic zone will look like the plane strain zone and on the surface it will look like the plane stress zone. Rather the plastic zones are much more complex. For very low loads the plastic zone in the mid-plane and on the surface have the plane strain "butterfly" shape. As the load increases the mid-plane plastic zone takes on a shape more like the "circular" plane stress zone, while the surface plastic zone resembles the plane strain shape. These analyses demonstrate that in the real (3D) world the plastic zones are configuration and load dependent and are three dimensional

deformed to a curve and that due to crack tip blunting the crack length has effectively increased although no material failure and hence no physical crack growth has occurred. This increase in crack length must be accounted for in the reduction of data to determine fracture toughness J_{IC}. Comparing the surface displacement field from the 3D FEM to 2D plane stress results, the study shows that 3D effects

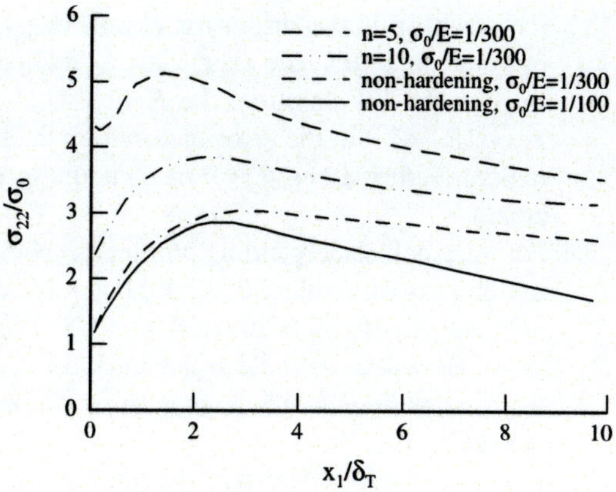

Fig. 7.38 Tensile stress ahead of plane strain crack. Effect of crack tip blunting is to reduce the stress in the crack tip region, $r < 3\delta_T$. Adapted from [24]

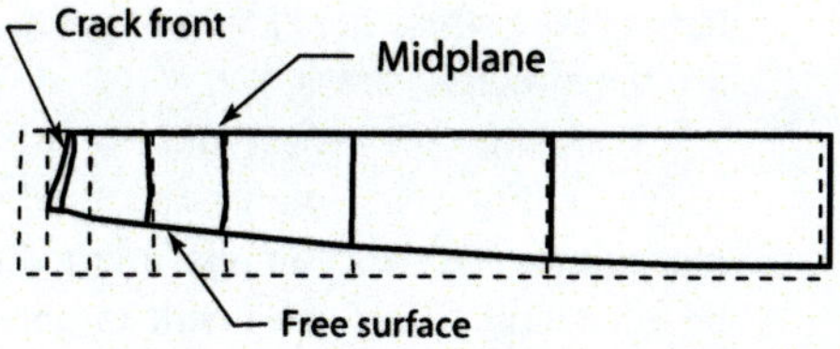

Fig. 7.39 Deformed and undeformed mesh for a 3D, large deformation analysis of a Mode-I crack in a power law hardening material with $n = 10$, adapted from [25]

can extend up to two plate thicknesses ahead of the crack for materials with strain hardening, and further in non-hardening materials.

7.7 Exercises

1. Repeat the analysis of the strip yield model for a semi-infinite crack loaded with stress intensity factor K_I. Use Eqs. (5.2) for the stress intensity factor analysis and (5.3) for the displacement. Plot the shape of the crack opening at the crack tip, i.e. in and behind the yield zone. Note the cusp-like shape of the displacement along the strip yield zone. Show that s for the semi-infinite crack case is the same as that for SSY in the finite crack. Explain this result.
2. For the strip yield zone problem of a finite crack in an infinite plate solve for and plot the stress at the crack tip by solving for ϕ' for all three problems superposing the results and using $\sigma_{22} = \mathrm{Re}\,\phi'$ along $z = x_1$.
3. Verify Eq. (7.42).
4. Work out the details and verify Eq. (7.28). Hint: write everything in indicial notation and note that $\partial a_{ij}/\partial a_{kn} = \delta_{ik}\delta_{jn}$.
5. Verify Eq. (7.29).
6. Verify Eq. (7.34).
7. Show that the plastic zone Eq. (7.41) satisfies the Irwin plastic zone criterion.
8. The elastic field for the stationary crack in elastic-perfectly plastic material under Mode-III loading could be written in complex variables as $\tau = \frac{K_{III}}{\sqrt{2\pi(z-R)}}$, where

the crack tip is the origin for z. (a) For $|z| > 2R$, expand τ in a power series of z and show that the elastic stress field consists of $r^{-1/2}, r^{-3/2}, r^{-5/2}$ terms. (b) Plot the full solution, the K field ($r^{-1/2}$ term) and $r^{-3/2}$ terms on $\theta = 0$, $x_1 > 2R$. (c) Plot the error incurred by including only the K field in the elastic region. (Either along $\theta = 0$ or contours in the entire region outside the plastic zone.)

9. For the small scale yielding problem, $\rho \neq 0$, shown in Fig. 7.7 taking both positive and negative values of λ_n use boundary collocation to calculate the full stress field and K_{III} for $\frac{\rho}{R} = \{0.01, 0.02, 0.05, 0.1\}$. Use $F(\theta) = \sin\theta$.
   (a) How does the hole at the crack tip affect the maximum stress?
   (b) How does K_{III} change with ρ/R? Can you give a physical reason for the change?
   (c) For $\rho/R = 0.02$ plot the ratio of the stress from the K_{III} field to the full field stress. What is the maximum error incurred by neglecting the more singular terms in the stress field? Where is the error minimized? For this problem you can use the effective stress, $|\tau|$.
   (d) How do the higher and lower order terms vary as the hole size increases?
   Notes: For very small holes the boundary collocation method may produce inaccurate results if you take a large number of negative λ_n terms since ρ^{λ_n} will be very large. This can result in an ill-conditioned matrix in which the matrix components differ by many orders of magnitude. You should be able to obtain good results with terms down to $r^{-5/2}$.

References

1. D.S. Dugdale, J. Mech. Phys. Solids **8**, 100 (1960)
2. J.W. Hutchinson, *A Course on Nonlinear Fracture Mechanics* (Department of Solid Mechanics, The Technical University of Denmark, 1979)
3. C.Y. Hui, A. Ruina, Int. J. Fract. **72**, 97 (1995)
4. A. Mendelson, *Plasticity: Theory and Application* (Krieger, Malabar, 1983)
5. J.R. Rice, in *Fracture an Advanced Treatise*, vol. 2, ed. by H. Liebowitz (Academic Press, New York, 1968), pp. 191–311, Chap. 3
6. A.D. Chitaley, F.A. McClintock, J. Mech. Phys. Solids **19**, 147 (1971)
7. W.R. Andrews, C.F. Shih, Elast.-Plast. Fract. **668**, 426–450 (1979)
8. J.W. Hutchinson, J. Mech. Phys. Solids **16**, 13 (1968)
9. J.R. Rice, G.F. Rosengren, J. Mech. Phys. Solids **16**, 1 (1968)
10. D. Carka, C.M. Landis, Journal of Applied Mechanics (2010, in press). doi:10.1115/1.4001748
11. A.P. Green, B.B. Hundy, J. Mech. Phys. Solids **4**, 128 (1956)
12. V. Kumar, M.D. German, C.F. Shih, Engineering approach for elastic-plastic fracture analysis. Tech. Rep. NP-1931, Electric Power Research Institute (1981)
13. N. Levy, P.V. Marcal, W.J. Ostergren, J.R. Rice, Int. J. Fract. Mech. **7**, 143 (1971)
14. R. Narasimhan, A.J. Rosakis, J. Mech. Phys. Solids **36**, 77 (1988)
15. C.F. Shih, J. Mech. Phys. Solids **29**(4), 305 (1981)
16. P. Ponte-Castaneda, J. Appl. Mech. **54**, 379 (1987)
17. R. Narasimhan, A.J. Rosakis, J.F. Hall, J. Appl. Mech. **54**, 838 (1987)
18. J.R. Rice, W.J. Drugan, T.L. Sham, in *Fracture Mechanics: Twelfth Conference*. Elastic-Plastic Fracture, vol. 700 (1980), pp. 189–221

19. J.R. Rice, in *Mechanics and Mechanisms of Crack Growth*, ed. by N.J. May, British Steel Corporations, Sheffield, 1975 (1975), pp. 14–39, chap. Elastic-plastic models for stable crack growth
20. R. Narasimhan, A.J. Rosakis, J. Appl. Mech. **57**, 607 (1990)
21. A.T. Zehnder, A.J. Rosakis, J. Appl. Mech. **57**, 618 (1990)
22. J.R. Rice, P.C. Paris, J.G. Merkle, in *Progress in Flaw Growth and Fracture Toughness Testing*. Elastic-Plastic Fracture, vol. 536 (American Society for Testing and Materials, 1973), pp. 231–245
23. T. Nakamura, D.M. Parks, in *Analytical, Numerical and Experimental Aspects of Three Dimensional Fracture Processes*, vol. AMD-91, ed. by A.J. Rosakis, K. Ravi-Chandar, Y. Rajapakse (American Society of Mechanical Engineers, 1988), pp. 227–238
24. R.M. McMeeking, J. Mech. Phys. Solids **25**, 357 (1977)
25. C.L. Hom, R.M. McMeeking, Int. J. Fract. **45**, 103 (1990)

Chapter 8
Elastic Plastic Fracture: Energy and Applications

Abstract The energy consumed in the growth of crack in an elastic-plastic material is mostly the work associated with plastic deformation around the crack. Methods for computing G and J in elastic-plastic fracture are discussed in this chapter. The limitations of a J based approach to predicting fracture are explored as are several other criteria for ductile fracture, including the J–Q theory, critical crack tip opening angle and the use of cohesive zone models.

8.1 Energy Flows

In an elastic-plastic material J and G are no longer equal even for a crack that grows straight ahead. However we have seen that J characterizes the amplitude of the crack tip stress and strain fields for hardening materials. Furthermore for stationary cracks and for cracks that grow by $\Delta a \ll r_p$ J is nearly path independent. Thus although $J \neq G$, J is still a useful concept.

8.1.1 When Does G = J?

Consider two cracked bodies, with identical loading and geometry as shown in Fig. 8.1. One body is made of material 1, a deformation theory i.e. nonlinear elastic, material. The other body is made of a material 2, an incremental, elastic-plastic material.

We have shown that for elastic materials $J = G$, for straight ahead crack growth, thus $J_1 = G_1$. However, in elastic-plastic materials the unloading that occurs during crack growth does not follow (in terms of stress vs. strain) the same path as loading and thus $J_2 \neq G_2$. Furthermore the J integral is not path independent. Nonetheless, up to the point where the crack begins to grow the deformation theory and incremental theory stress and strain fields may be similar enough that one can make the *approximation* that $J_2 \approx J_1$, and hence that $J_2 = G_1$, i.e. the far field J integral prior to crack growth for the incremental theory material is approximately equal to the energy release rate that you would have if the crack were to grow in the matching deformation theory material.

A.T. Zehnder, *Fracture Mechanics*,
Lecture Notes in Applied and Computational Mechanics 62,
DOI 10.1007/978-94-007-2595-9_8,

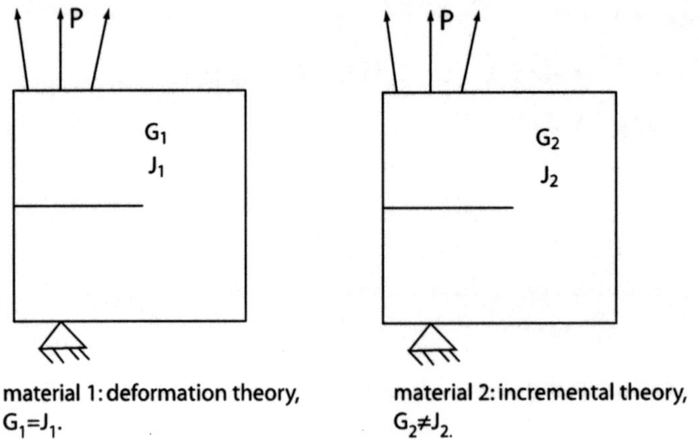

Fig. 8.1 Energy release rate and J for nonlinear elastic and elastic-plastic materials with same loading behavior

What good is this result? From Eqs. (3.20) and (3.22) we know how to calculate G for a nonlinear elastic material in terms of the changes in the load-displacement record with respect to crack length. Thus this method may be applied directly to the computation of J in elastic-plastic materials, either by performing experiments with samples of different initial crack lengths, or through the use of analytical or computational approximations to the functions $P(a,q)$ and $q(a,P)$.

8.1.2 General Treatment of Crack Tip Contour Integrals

The idea of energy release rate can be generalized. As discussed in refs. [1, 2] consider first a general balance statement between vector field ϕ_j and scalar field ψ,

$$\phi_{j,j} = \dot{\psi}.$$

In heat transfer ϕ_j is the heat flux and ψ is the temperature. An integral balance statement can be formed from the local balance,

$$\int_{\mathscr{V}} \phi_{j,j}\, dV = \int_{\mathscr{V}} \dot{\psi}\, dV. \tag{8.1}$$

Recall the transport theorem

$$\frac{d}{dt}\int_{\mathscr{V}} f\, dV = \int_{\mathscr{V}} \dot{f}\, dV + \int_{\mathscr{S}} v_i m_i\, dS,$$

where f is any field, v_i is the velocity of the surface $\mathscr{S}$ of $\mathscr{V}$ and m_i is the unit normal to $\mathscr{S}$. Applying the transport theorem to the right hand side and divergence

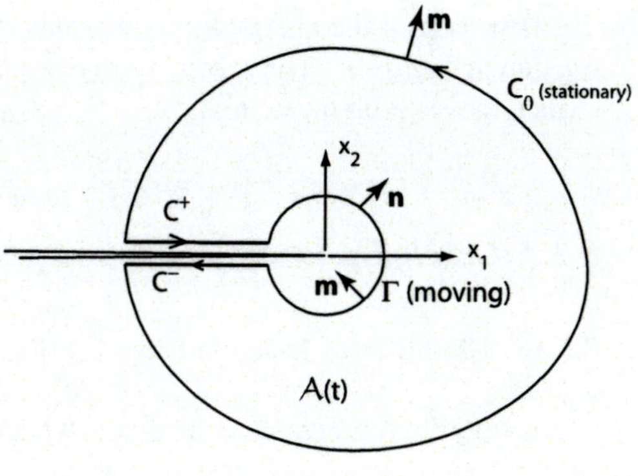

Fig. 8.2 Contours for generalized contour integral

theorem to the left hand side of Eq. (8.1), the balance statement for a moving region can be written as

$$\int_{\mathscr{S}} \phi_j m_j \, dS = \frac{d}{dt} \int_{\mathscr{V}} \psi \, dV - \int_{\mathscr{S}} \psi v_i m_i dS. \tag{8.2}$$

Let us consider a crack moving in the x_1 direction at speed v. Surround the crack tip with the closed area $A(t)$ shown in Fig. 8.2. A contour γ moves with the crack tip, while the outer contour C_0 is stationary. Specializing equation (8.2) for the 2D domain $A(t)$ and noting that on Γ $\mathbf{n} = -\mathbf{m}$, the terms of the equation are

$$\int_{\mathscr{S}} \phi_j m_j \, dS = \int_{C_0} \phi_j m_j \, d\Gamma - \int_{\Gamma} \phi_j n_j \, d\Gamma \quad \text{assuming } \phi_j m_j = 0 \text{ on } C^+, C^-,$$

$$\frac{d}{dt} \int_{\mathscr{V}} \psi \, dV = \frac{d}{dt} \int_{A(t)} \psi \, dA,$$

$$\int_{\mathscr{S}} \psi v_j m_j \, dS = -\int_{\Gamma} \psi v_j n_j \, d\Gamma = -\int_{\Gamma} \psi v n_1 \, d\Gamma.$$

Substituting the above into Eq. (8.2) and re-arranging

$$\int_{C_0} \phi_j m_j \, d\Gamma = \frac{d}{dt} \int_{A(t)} \psi \, dA + \int_{\Gamma} (\phi_j + \psi v \delta_{1j}) n_j \, d\Gamma. \tag{8.3}$$

The interpretation of the above is straight forward. The term on the LHS represents the rate of energy input to the region surrounded by C_0. The first term on the RHS is the rate of increase of internal energy in $A(t)$, while the second term represents the flux of energy to the crack tip.

Specializing the above to solid mechanics, $\psi = W + T$, where W is the stress work density, T is the kinetic energy density, $\dot{W} = \sigma_{ij} \dot{\gamma}_{ij}$ and $\dot{T} = \rho \ddot{u}_i \dot{u}_i$ and $\phi_j = \sigma_{ij} \dot{u}_i$. It is easily verified that the above substitutions satisfy the balance, Eq. (8.1), thus the second term on the RHS of Eq. (8.3) representing the flux of energy to the crack tip is

$$F(\Gamma) = \int_{\Gamma} \left(\sigma_{ij} n_j \dot{u}_i + (W + T) v n_1 \right) d\Gamma. \tag{8.4}$$

The first term of the energy flux represents the work done on material inside Γ by traction acting on Γ. The second term represents the transport of energy through Γ (which moves with the crack tip).

8.1.3 Crack Tip Energy Flux Integral

8.1.3.1 Global Path Independence for Steady State Crack Growth

The energy flux integral is generally *not* path independent, however for the special case of steady state crack growth F is path independent. Under steady state conditions $\dot{f} = -v\partial f/\partial x_1$ and

$$F(\Gamma) = v\int_\Gamma \big(-\sigma_{ij}u_{i,1}n_j + (W+T)n_1\big)\,d\Gamma.$$

To show path independence consider a closed contour such as that shown in Fig. 3.11 (but with Γ moving with the crack tip). Let A be the region inside the closed contour. Then writing the integral over the closed contour and applying the divergence theorem,

$$\begin{aligned}
F(\Gamma_1 - \Gamma_2 + \Gamma^{(+)} + \Gamma^{(-)}) &= v\oint_\Gamma \big(-\sigma_{ij}u_{i,1}n_j + (W+T)n_1\big)\,d\Gamma \\
&= v\int_A \big(W_{,1} + T_{,1} - (\sigma_{ij}u_{i,1})_{,j}\big)\,dA \\
&= v\int_A (W_{,1} + T_{,1} - \sigma_{ij,j}u_{i,1} - \sigma_{ij}u_{i,1j})\,dA.
\end{aligned}$$

Using the steady state condition that $f_{,1} = -\dot{f}/v$ and using the equation of motion, $\sigma_{ij,j} = \rho\ddot{u}_i$, we have

$$\begin{aligned}
F(\Gamma_1 - \Gamma_2 + \Gamma^{(+)} + \Gamma^{(-)}) &= v\int_A \left(-\frac{\dot{W}}{v} - \frac{\dot{T}}{v} - \rho\ddot{u}_i u_{i,1} - \sigma_{ij}\gamma_{ij,1}\right) dA \\
&= v\int_A \left(-\frac{\dot{W}}{v} - \frac{\dot{T}}{v} + \rho\ddot{u}_i\frac{\dot{u}_i}{v} + \sigma_{ij}\frac{\dot{\gamma}_{ij}}{v}\right) dA \\
&= 0 \quad \text{using } \dot{W} = \sigma_{ij}\dot{\gamma}_{ij} \text{ and } \dot{T} = \rho\ddot{u}_i\dot{u}_i.
\end{aligned}$$

On the contours $\Gamma^{(+)}$ and $\Gamma^{(-)}$, $n_1 = 0$ and $\sigma_{ij}n_j = 0$ thus $F(\Gamma_1 - \Gamma_2 + \Gamma^{(+)} + \Gamma^{(-)}) = F(\Gamma_1) - F(\Gamma_2) = 0$ and hence $F(\Gamma_1) = F(\Gamma_2)$ proving global path independence in the steady state case.

Since the velocity factors out of the integral in the steady state case, let us define

$$I(\Gamma) \equiv \frac{F(\Gamma)}{v}. \tag{8.5}$$

8.1.3.2 Energy Flux as $\Gamma \to 0$

We consider the case where Γ is an infinitesimal contour surrounding a moving crack tip. In this case the energy flux integral can be interpreted as the crack tip energy release rate, denoted as G_{tip} to distinguish it from the definition of G for elastic materials,

$$G_{tip} = \lim_{\Gamma\to 0} \frac{1}{v} F(\Gamma) = \lim_{\Gamma\to 0} \frac{1}{v} \int_{\Gamma} \big(\sigma_{ij} n_j \dot{u}_i + (W+T) v n_1\big)\, d\Gamma. \qquad (8.6)$$

Any singular field variable can be shown to satisfy the steady state condition as $r \to 0$. Consider $f = f(x_1 - a(t), x_2, t)$, $\dot{f} \equiv Df/Dt = -\dot{a}\partial f/\partial x_1 + \partial f/\partial t$. If f is singular then as $r \to 0$ the first term will be more singular than the second and will dominate as $r \to 0$. Thus for singular fields $\dot{f} \to -f_{,1} v$ where $v = \dot{a}$.

Applying the condition that the fields are locally steady state, and hence that F is locally path independent, we can choose Γ to be a circular contour of radius r and write G_{tip} as

$$G_{tip} = \lim_{\Gamma\to 0} I(\Gamma) = \lim_{r\to 0} \int_{-\pi}^{\pi} \big(-\sigma_{ij} u_{i,1} + (W+T)\delta_{1j}\big) n_j r\, d\theta. \qquad (8.7)$$

If $(-\sigma_{ij} u_{i,1} + (W+T)\delta_{1j}) n_j = A(\theta)/r + o(1/r)$ then G_{tip} will integrate to a finite, non-zero value. This is the case for cracks in linearly elastic materials and is true for stationary cracks in elastic-plastic materials, meaning that G_{tip} would be finite for crack growth in deformation theory materials. In addition G_{tip} is non-zero for crack growth in rate-dependent (viscoplastic) materials [3].

If $(-\sigma_{ij} u_{i,1} + (W+T)\delta_{1j}) n_j = o(1/r)$, then $G_{tip} = 0$. This is the case for a growing crack in an elastic-plastic material where $\sigma \cdot \gamma \sim \sigma_0 \ln r$. This is an interesting result. It says that the energy release rate for crack growth in elastic-plastic materials is zero. Where does the energy go?

8.1.3.3 Energy Flux for Γ Outside Plastic Zone

Consider the case where Γ is grown to a contour that is large in comparison to the active plastic zone (the region in which $d\gamma_{ij}^p \neq 0$). In this case Freund and Hutchinson [3] show that

$$\lim_{\Gamma\to\infty} I = G - \lim_{x_1\to -\infty} \int_{-h}^{h} U^* \, dx_2, \qquad (8.8)$$

where $U^*(x_2)$ is the stress work density locked in the plastic wake which spans from $-h$ to $+h$ along the crack line as shown in Fig. 8.3 and $G = K_I^2/E'$ (for quasi-static crack growth) is the energy release rate coming from the K field.

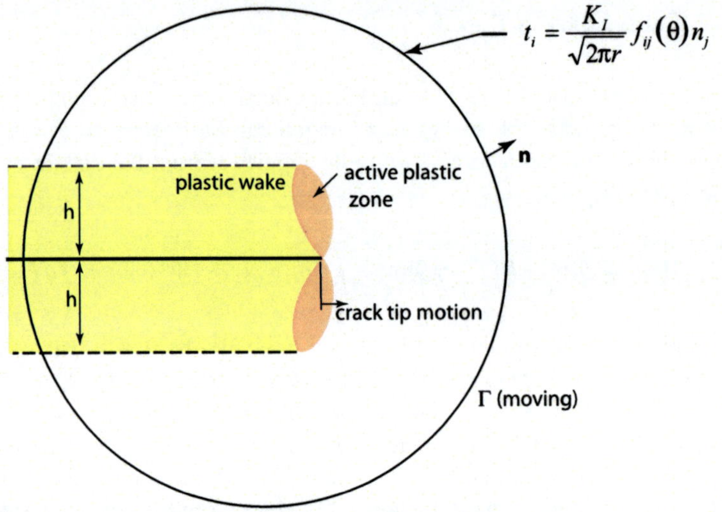

Fig. 8.3 Growing crack in elastic plastic material under SSY conditions

Breaking up U^* into elastic and plastic parts, $U^* = U^{*e} + U^{*p}$ it is shown that the plastic part is

$$\lim_{x_1 \to -\infty} \int_{-h}^{h} U^{*p}\, dx_2 = \frac{1}{v} \int_{A_p} \sigma_{ij} \dot{\gamma}_{ij}^{p}\, dA,$$

where A_p is the active plastic zone [3]. Using global path independence and substituting for U^{*p} we have

$$I = G_{tip} = G - \lim_{x_1 \to -\infty} \int_{-h}^{h} U^{*e}\, dx_2 - \frac{1}{v} \int_{A_p} \sigma_{ij} \dot{\gamma}_{ij}^{p}\, dA.$$

Re-arranging,

$$G = G_{tip} + \frac{1}{v} \int_{A_p} \sigma_{ij} \dot{\gamma}_{ij}^{p}\, dA + \int_{-h}^{h} U^{*e}\, dx_2, \tag{8.9}$$

where U^{*e} is the elastic strain energy locked in the plastic wake. The above says that the global energy release rate is partitioned into a component that flows to the crack tip, the integrated plastic work rate ahead of the crack and the flow of elastic energy into the wake. The elastic energy in the wake is generally much smaller than the plastic dissipation and is hence ignored in the following.

For a rate independent elastic-plastic material $G_{tip} = 0$, i.e. there is no energy flowing to the crack tip! Equation (8.9) above suggests that in elastic-plastic fracture the energy required to drive the crack goes not into new fracture surface energy but into the energy of inelastic deformation in the inelastic zone surrounding the crack tip. Even if $G_{tip} \neq 0$ (the case for dynamically growing cracks in rate dependent plastic materials) the energy required to break the bonds of the material at the crack tip is a small fraction of the total energy of failure in all but the most brittle materials. Further if the elastic portion of the energy locked in the wake is small (as it would be in most cases) then for elastic-plastic materials

$$G = \frac{1}{v} \int_{A_p} \sigma_{ij} \dot{\gamma}_{ij}^{p} \, dA, \tag{8.10}$$

where A_p is the area of the active plastic zone. Thus, G, the energy consumed per unit crack advance, is equal to the rate of plastic work dissipated around the crack tip.

For elastic materials $U^{*e} = 0$ and $\dot{\gamma}_{ij}^{p} = 0$, thus $G = G_{tip}$ in agreement with path independence of the J integral shown in Sect. 3.7.

Although the above analysis holds strictly only for steady-state crack growth it is expected to be a good approximation of the actual energy flows as long as the changes with respect to crack growth are modest. As an example demonstrating that virtually all of the energy release rate for crack growth in elastic-plastic materials is consumed as plastic work, consider the example below in which imaging of the temperature rise due to thermal dissipation of plastic work at the crack tip is used to calculate G.

8.1.3.4 Thermal Field Visualization of Energy Flow

Bhalla et al. [4] illustrate the above result using high speed thermal imaging. Most of the energy of plastic deformation is dissipated thermally with only a small portion remaining locked into the material in the form of elastic strain energy due to incompatible deformations and interactions of dislocations and dislocation clusters. Thus, to a good approximation, plastic work is a distributed heat source $\beta \sigma_{ij} \dot{\gamma}_{ij}^{p}$ where β, the fraction of plastic work thermally dissipated is typically around 0.8 to 0.95. With this heat source, ignoring convective and radiation heat losses and ignoring the thermoelastic effect (all of these are shown to be small) the heat conduction equation is

$$\beta \sigma_{ij} \dot{\gamma}_{ij}^{p} + k \nabla^2 T = \rho c \dot{T}, \tag{8.11}$$

where T is temperature, k is thermal conductivity, ρ is density and c specific heat. Solving for the plastic work rate, $\sigma_{ij} \dot{\gamma}_{ij}^{p}$, and substituting into Eq. (8.10) we have

$$G = \frac{1}{\beta \dot{a}} \int_{A_p} (\rho c \dot{T} - k \nabla^2 T) \, dA.$$

Applying the divergence theorem and noting that if the field of view of the thermal image is taken large enough no heat conducts out of the field of view, it is shown that

$$G = \frac{1}{\beta \dot{a}} \frac{d}{dt} \int_{FOV} \rho c T \, dA, \tag{8.12}$$

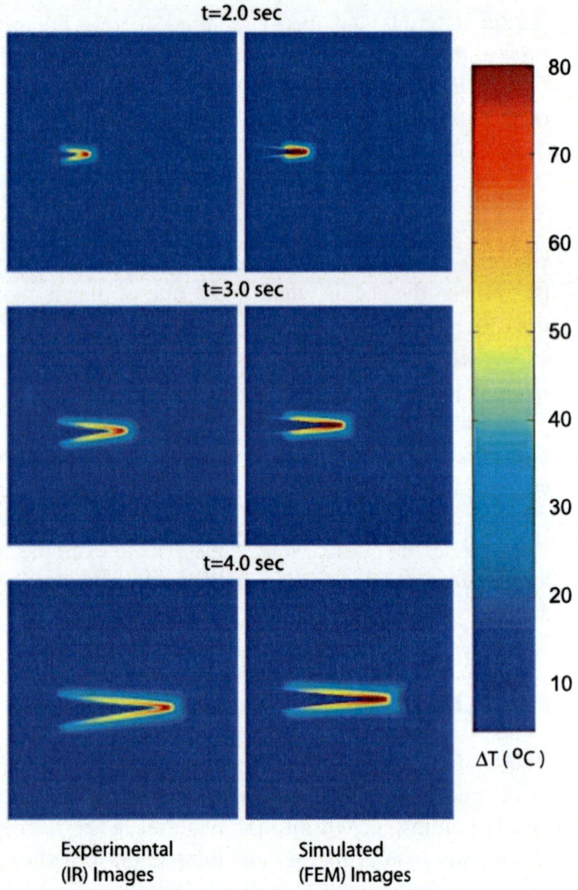

Fig. 8.4 Comparison of the temperature field history from thermal imaging, (**a**), and Finite Element Analysis, (**b**), for a test performed at a loading rate of 5.1 mm/sec and imaged in the 172 mm × 172 mm field of view. Reprinted from Engineering Fracture Mechanics, Vol. 70, K.S. Bhalla, A.T. Zehnder and X. Han, "Thermomechanics of slow stable crack growth: closing the loop between experiments and computational modeling," 2439–2458, copyright (2003), with permission from Elsevier, [4]

where *FOV* is the entire field of view of the thermal camera.

Single edge notched samples (152 mm wide, 237 mm long and 0.8 mm thick, with 30 mm initial crack) of annealed 302 SS were loaded at constant displacement rates and imaged with an infrared camera as the crack grew. A selection of images obtained from the experiments is shown in Fig. 8.4. Temperature increases, ΔT, of up to 80°C were observed at the crack tip. Using values for β measured for this material the temperature field was simulated using a finite element method in which the plastic work rate is computed and used as input to a transient thermal computation. The resulting simulated thermal field, also shown in the figure shows good agreement with the measured field indicating that the model for thermal dissipation of plastic work is valid.

Using an average value of $\beta = 0.85$, G was computed using Eq. (8.12). The resulting energy flow to the crack is plotted in Fig. 8.5 along with the values computed directly from the finite element simulations. The good agreement between the two results illustrates the validity of Eq. (8.10) above. Note that the crack grows with almost constant energy flux until the crack nears the edge of the test specimen and

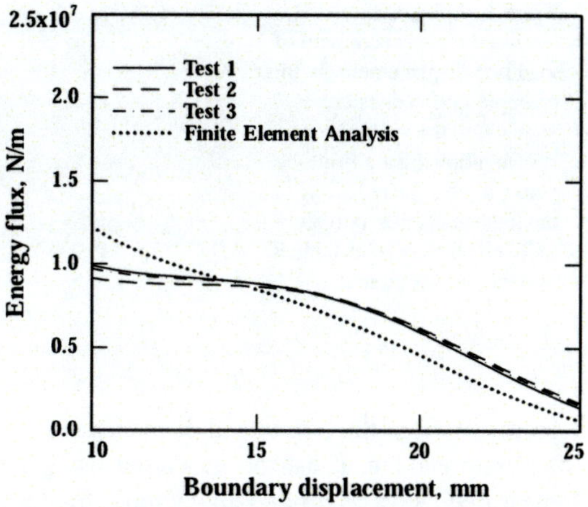

Fig. 8.5 Energy flux to the crack computed from infrared images and calculated from FEA vs. stretching displacement applied to the single edge notched sample. (Tests 1, 2 loaded at displacement rate of 5.1 mm/sec and Test 3 at rate of 16.9 mm/sec.) From [4]

progressively less energy is required to drive the crack through the fully yielded remaining ligament.

8.2 Fracture Toughness Testing for Elastic-Plastic Materials

Elastic-plastic fracture toughness is typically characterized in terms of critical values of J. Test methods for toughness measurements are described in detail in ASTM E1820 [5], which combines methods for the measurement of J_C, (the plane strain value of J prior to significant crack growth), J_{IC}, (the value of J near the onset of stable crack growth) and J_R the J vs. Δa resistance curve for stable crack growth. The tests are based on the measurement of load and load-line displacement to determine fracture energy and on compliance to determine crack length.

8.2.1 Samples and Equipment

Either compact tension or single edge notched bending specimens may be used. As with E-399 the samples should be notched and precracked by fatigue loading. Side grooves may be cut in the sample to ensure straight ahead crack growth. The samples are to be sized so that the uncracked ligament length $W - a > 25(J_{IC}/\sigma_Y)$ and the thickness, $B > 25(J_{IC}/\sigma_Y)$, where σ_Y is the average of the yield strength and ultimate tensile strengths,

$$\sigma_Y \equiv \frac{\sigma_0 + \sigma_{TS}}{2}$$

and σ_{TS} is the ultimate tensile strength.

For the SENB specimen the load line displacement, v_L and crack mouth opening displacement, (CMOD) v_c must be measured. Load line displacement can be

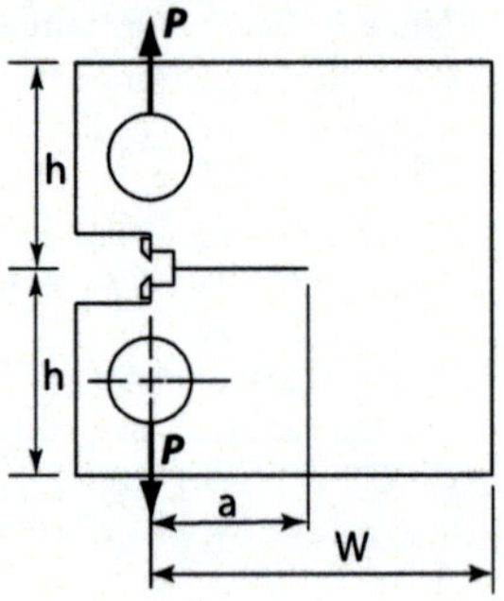

Fig. 8.6 CT specimen modified for measurement of load line displacement. In this example knife edges are attached to the specimen as shown, allowing a clip-in gauge to be attached along the load-line. Alternatively, knife edges may be machined into the test specimen

measured from the cross-head displacement of the testing machine, or if this is not accurate enough, it can be measured using an LVDT or other gauge. CMOD is measured using a clip-in gauge similar to that used for K_{IC} testing, (see Fig. 6.2) although the range of the gauge may need to be larger due to plastic deformation.

The CT specimen must be modified in order to measure the load line displacement. An example is shown in Fig. 8.6 where the crack mouth width is increased and knife edges are attached to the specimen. A displacement gauge clipped in to these knife edges is used to measure the load line displacement.

8.2.2 Procedure and Data Reduction

Two test methods, a basic and R-curve method are discussed. In the basic method multiple specimens are used to determine J_{IC}. In the R-curve method loading-unloading cycles are used to determine $J(\Delta a)$ and from this J_{IC} and the J–R curve. In the basic method, only the load and load point displacement need to be measured. In the R-curve method the load, P, load-point displacement, v_L and crack mouth opening displacement, v_C, must be measured. Here only the R-curve is discussed since it is the more general of the two approaches.

8.2.2.1 Test Procedure

The first step is to cycle the specimen three times from 0.5 to 1.0 the maximum load used in precracking. Estimate the initial crack length, a_0 from the crack mouth opening compliance measured during these cycles. Proceed with the testing, unloading at load-line displacement intervals of approximately $0.005W$. To avoid reverse plastic loading at the crack tip, do not unload below 50% of the maximum load reached. At least 8 load-unload cycles should be performed in order to obtain sufficient J vs. Δa data. After the final loading step, unload the sample to zero and mark the final crack length either by heat tinting and or by fatigue loading. Once the sample has been marked break open (may cool steel samples to ensure brittle fracture) and measure the final crack length.

8.2.2.2 Data Reduction

The load-deflection curve obtained with this procedure will look something like the plot in Fig. 8.7. For each cycle determine the compliance, C, and then determine the crack length a_i for each cycle by inverting the $C(a)$ relation. (Formulas are given in [5] for several standard specimen geometries.) For the SENB specimen a_i is calculated from

$$\frac{a_i}{W} = 0.999748 - 3.9504u + 2.9821u^2 - 3.21408u^3 + 51.5156u^4 - 113.031u^5,$$

$$u \equiv \frac{1}{\left[\frac{B_e W E C_i}{S/4}\right]^{1/2} + 1},$$

$$B_e = B - \frac{(B - B_N)^2}{B}, \tag{8.13}$$

where B_N is the width of the sample after notching with side-grooves (if notched) and $C_i = \Delta v_c / \Delta P$ for a cycle. You should now have a table of P and a values.

To determine J use

$$J = J_{el} + J_{pl}, \tag{8.14}$$

$$J_{el} = \frac{K_I^2}{E'},$$

where K_I is calculated from the SENB formula in Table 5.1 with B replaced by $\sqrt{B B_N}$ if sample contains side-grooves. The second term, J_{pl} is calculated using the load and load-line displacement, v_L data. First the plastic part of v_L is calculated using the measured values of δ and P and the calculated values of a:

$$v_L^{pl} = v_L - P C_L(a), \tag{8.15}$$

where C_L is the load-line compliance that you would have if all displacement were elastic. For the SENB specimen, this may be calculated from the formula for v_L in Table 5.1 with B replaced by B_e. For each of the pairs of P_i, a_i data calculate J_{pl} from

$$(J_{pl})_i = (J_{pl})_{i-1} + \frac{1.9}{b_{i-1}} \frac{(\Delta A_{pl})_i}{B_N} \left[1 - 0.9 \frac{\Delta a}{b_{i-1}}\right],$$
$$\Delta a = a_i - a_{i-1}, \tag{8.16}$$

where $b = W - a$ is the uncracked ligament length and $(\Delta A_{pl})_i$ is the area under the P, v_L^{pl} curve from load P_{i-1} to load P_i, i.e.

$$(\Delta A_{pl})_i = \int_{(v_L^{pl})_{i-1}}^{(v_L^{pl})_i} P \, dv_L^{pl}.$$

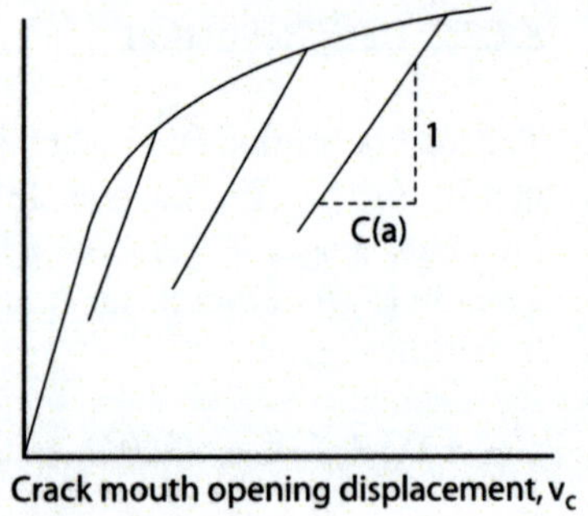

Fig. 8.7 Schematic load vs. crack mouth opening displacement plot for R-curve method. Only three cycles are shown. At least 8 load-unload cycles should be used to obtain sufficient J vs. Δa data

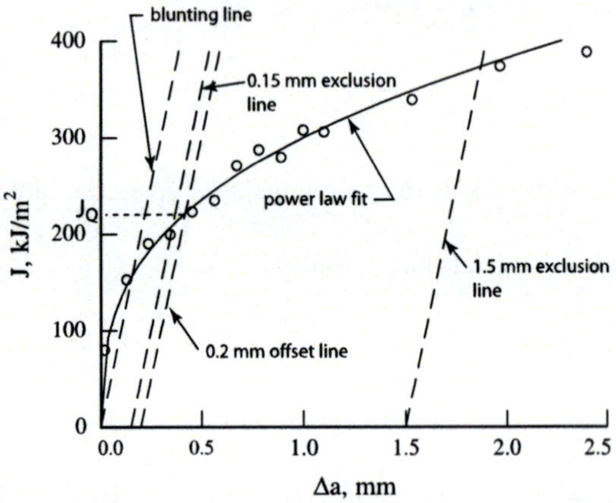

Fig. 8.8 J vs. Δa data and construction lines for data qualification. Adapted from [5]

8.2.2.3 Validation of Results

To determine J_{IC} and to determine the valid range of the J–R curve, plot the J vs. Δa data obtained from the above procedures.

The results are sensitive to the initial crack length, a_0. To ensure that the initial crack length is accurate, the standard calls for a_0 to be adjusted by fitting a third order polynomial to the J vs. Δa data and extracting an adjusted initial crack length called a_{0Q},

$$a = a_{0Q} + \frac{J}{2\sigma_Y} + BJ^2 + CJ^3.$$

If the a_{0Q} value differs from the optically measured a_0 by more than $0.01W$ the data are invalid. If not, then from your set of J and a values, plot J vs. $\Delta a = a - a_{0Q}$. An example is shown in Fig. 8.8.

Even with no physical crack growth, the crack will appear to have grown due to blunting as seen in Fig. 7.39 where the crack front after deformation but prior to crack growth is advanced relative to the initial crack line. To account for this, first draw the "blunting line" given by

$$J = 2\sigma_Y \Delta a$$

on the J, Δa plot. The difference of Δa and the blunting line represents the physical crack growth. Next draw exclusion lines parallel to the blunting line but offset by 0.15 and 1.5 mm. Draw a 0.2 mm offset line parallel to the blunting line but shifted by 0.2 mm. Draw a fourth line (not shown in the figure) with a 0.5 mm offset. You must have at least one data point between the 0.15 and 0.5 mm lines and at least one between the 0.5 and 1.5 mm lines and at least 4 between the 0.15 and 1.5 mm lines. Assuming that these conditions are met, fit a power law, $J = C_1 \Delta a^{C_2}$ to the data that lie in the exclusion zone bounded by the 0.15 and 1.5 mm exclusion lines. The power law can be found by linear fitting of the data to

$$\ln J = \ln C_1 + C_2 \ln(\Delta a).$$

Plot the power law relationship. Where it intersects the 0.2 mm offset line, $J = J_Q$, the provisional fracture toughness and $\Delta a = \Delta a_Q$. Then $J = J_Q$ if the plane strain constraint conditions are met,

$$\{B, b_0\} > 25(J_Q/\sigma_o)^2$$

and if

$$\frac{dJ}{da} < \sigma_Y \quad \text{at } \Delta a = \Delta a_Q.$$

The J–R curve is defined as the J vs. Δa data over a broader region bounded by $\Delta a_{max} = 0.25 b_0$, and J_{max} given by the smaller of $b\sigma_Y/20$ or $B\sigma_Y/20$.

8.2.3 Examples of J–R Data

Examples of J–R data are shown in Figs. 8.9(a) and 8.9(b). The results show that the side-grooved sample has lowest J_{IC} and dJ/da due to suppression of plane-stress deformation on the free surfaces of the sample. Removal of the side grooves (100 mm sample) results in higher J. Thinner samples (25 and 62.5 mm thick) have higher J due to increased percentage of plane stress deformation. Figure 8.9(b) shows that crack length has some effect on J_{IC} and dJ/da. For 100 mm thick, 12.5% side-grooved samples with $W = 203$ mm values of J are seen to decrease slightly with increasing a. This is likely due to changes in constraint.

8.3 Calculating J and Other Ductile Fracture Parameters

To apply fracture toughness data to the prediction of fracture accurate means for the calculation of J must be available. Here I will briefly (and incompletely) discuss computational, analytical and handbook methods.

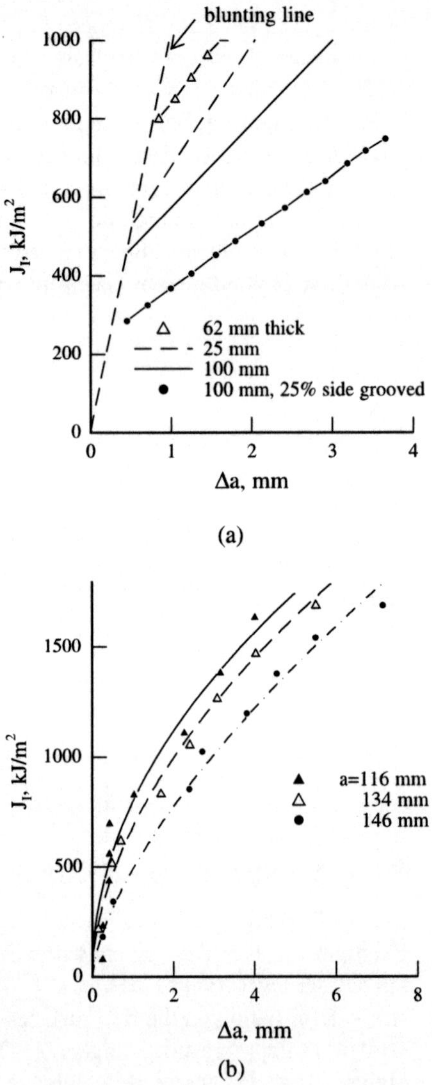

Fig. 8.9 *J–R* curve data for A533-B steel at 93°C, adapted from [6]. (**a**) Effect of thickness and side grooves. Side-grooved sample has lowest J_{IC} and dJ/da due to suppression of plane-stress deformation on the free surfaces of the sample. Removal of the side grooves (100 mm sample) results in higher J. Thinner samples (25 and 62.5 mm thick) have higher J due to increased percentage of plane stress deformation. (**b**) Crack length has some effect on J_{IC} and dJ/da. For 100 mm thick, 12.5% side-grooved samples with $W = 203$ mm values of J are seen to decrease slightly with increasing a. This is likely due to changes in constraint

8.3.1 Computational Methods

A very general approach would be to use the finite element method in conjunction with nonlinear constitutive models and calculate J using the domain integral. As with elastic fracture problems accuracy will be improved by incorporating the crack tip strain singularity into the computational model.

The HRR and Mode-III elastic-plastic analyses show that at the crack tip the strains will vary as r^{-s}, where $s = n/(n+1)$ ranges from 0.5 to 1. Given that most metals will have $n > 3$, the typical range is $0.7 < s < 1$, i.e. the strains are close to $1/r$ singular. Thus, although special crack tip elements for various values of n

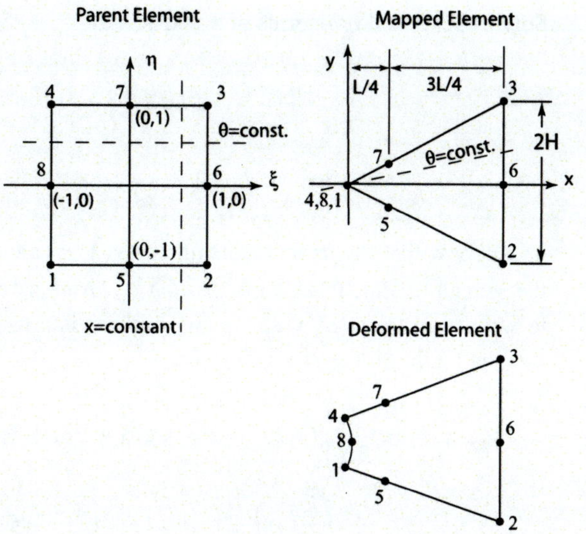

Fig. 8.10 Collapsed quadrilateral element for elastic-plastic fracture analysis. Lines of constant ξ in parent element map to constant x in mapped element. Lines of constant η map to constant θ. Adapted from [8]

have been developed [9, 10], it is common practice in elastic-plastic fracture to use crack tip elements that provide a $1/r$ singular strain field. Such elements were first introduced by Levy et al. [11].

Here we will follow the analysis of Barsoum [8]. As with the elastic singular elements we take as a starting point an eight-node rectangular isoparametric element and collapse the nodes 4, 8, 1 along $\xi = -1$ to a single point, see Fig. 8.10. In contrast with the elastic quarter point element do not tie nodes 4,8,1 together, i.e. allow them to move independently so that the crack can have a finite crack tip opening as shown in Fig. 8.10.

As shown in Fig. 8.10 the nodal coordinates are

$$
\begin{aligned}
x_1 = x_4 = x_8 &= 0, \\
x_2 = x_6 = x_3 &= L, \\
x_5 = x_7 &= L/4, \\
y_1 = y_8 = y_4 = y_6 &= 0, \\
y_5 = -y_7 &= -H/4, \\
y_2 = -y_3 &= -H.
\end{aligned}
$$

Using the interpolation for the coordinates, Eqs. (5.17) with the 8 node element shape functions given by Eqs. (5.19) and simplifying yields

$$
x = \frac{L}{4}(1+\xi)^2,
$$

$$
y = \frac{H}{4}\eta(1+\xi)^2.
$$

Solving for ξ and η in terms of x and y,

$$\xi = 2\sqrt{\frac{x}{L}} - 1, \qquad \eta = \frac{L}{H}\frac{y}{x}. \tag{8.17}$$

To show that the strains are singular as $1/r$ calculate the v displacement using the second of Eqs. (5.18), $v(\xi,\eta) = \sum_{i=1}^{8} N_i(\xi,\eta)v_i$. Then substitute Eqs. (8.17) to write v in terms of x and y and differentiate to calculate the strain $\gamma_{yy} = \frac{\partial v}{\partial y}$. Evaluating the strain on $y = 0$ yields

$$\gamma_{yy}(x,0) = \frac{1}{2H}\Big[(-2v_1 - 2v_2 + 2v_3 + 2v_4 + 4v_5 - 4v_7) + \sqrt{\frac{L}{x}}(3v_1 + v_2 - v_3 - 3v_4 - 4v_5 + 4v_7) + \frac{L}{x}(v_4 - v_1)\Big]. \tag{8.18}$$

Thus it is seen that the strain consists of constant, $1/\sqrt{r}$ and $1/r$ terms. If the crack tip nodes are tied together then $v_1 = v_4 = v_8$ and the $1/r$ singular term is suppressed, reverting to the elastic singular element. Using a more general calculation, Barsoum [8] shows that along *any* $\theta =$ constant line the element in Fig. 8.10 will have strains with $1/r$, $1/\sqrt{r}$ and constant terms. As Levy et al. discuss [11] the $1/r$ singularity can be obtained by collapsing one complete side of nodes to a point for any rectangular element, including linear (4 node) elements. For non-hardening materials better accuracy may be obtained by keeping the nodes at the midpoint [7], suppressing the $1/\sqrt{r}$ singularity associated with the elastic field.

8.3.2 *J* Result Used in ASTM Standard J_{IC} Test

J can be calculated analytically in special cases. One example is the result used for the basic method of the standard for J_{IC} testing [5] of SENB specimens.

Recall from Sect. 8.1.1 that prior to crack growth $J = G$ where G is the energy release that you would have if the crack grew in an identical body made of a fictitious nonlinear elastic material whose stress strain curve during is the same as the stress-train curve of the elastic-plastic material during loading. Recall as well from Eq. (3.22) that (replacing the generalized displacement q with v_L and the crack surface area s by Ba,

$$J = G = \frac{1}{B}\int_0^P \frac{\partial v_L}{\partial a}\, dP. \tag{8.19}$$

If we could determine $v_L(P,a)$ we could find J.

By dimensional analysis of the SENB specimen

$$\frac{v_L}{S} = f\left(\frac{PS}{EBb^2}, \frac{\sigma_0}{E}, \nu, \frac{S}{B}, \frac{W}{b}\right),$$

where $b = W - a$ is the uncracked ligament length and the other dimensions are as given in Table 5.1. This nondimensionalization is consistent with beam theory. Let $f_{,n}$ be the derivative of f with respect to the nth argument. Then

$$\frac{\partial v_L}{\partial a} = -\frac{\partial v_L}{\partial b} = S\left\{f_{,1}\frac{2PS}{EBb^3} + f_{,4}\frac{S}{b^2} + f_{,5}\frac{W}{b^2}\right\}. \tag{8.20}$$

Note that

$$\frac{\partial v_L}{\partial P} = Sf_{,1}\frac{S}{EBb^2},$$

thus

$$f_{,1} = \frac{\partial v_L}{\partial P}\frac{EBb^2}{S^2}. \tag{8.21}$$

Combining Eqs. (8.21) and (8.20)

$$-\frac{\partial v_L}{\partial b} = \frac{2P}{b}\frac{\partial v_L}{\partial P} + f_{,4}\frac{S^2}{b^2} + f_{,5}\frac{WS}{b^2}.$$

Applying Eq. (8.19)

$$\begin{aligned} J &= \frac{1}{B}\int_0^P -\frac{\partial v_L}{\partial b}\,dP \\ &= \frac{2}{Bb}\int_0^{P(v_L)} P\frac{\partial v_L}{\partial P}\,dP + \int_0^P f_{,4}\frac{S^2}{b^2}\,dP + \int_0^P f_{,5}\frac{WS}{b^2}\,dP \\ &= \frac{2}{Bb}\int_0^{v_L} P(v_L)\,dv_L + I_4 + I_5. \end{aligned}$$

Under certain conditions I_4 and I_5 may be neglected and

$$J = \frac{2}{Bb}\int_0^{v_L} P(v_L)\,dv_L. \tag{8.22}$$

Note that the expression used for the R-curve method in [5] is essentially a discreet version of this integral with a correction for the change in b due to crack growth.

In the following we examine two extreme cases, rigid plastic and linearly elastic material behavior.

8.3.2.1 Rigid Plastic Material

If the material is rigid plastic then [12] once the sample is fully yielded and P reaches the limit load,

$$P = P_L = 4\frac{0.35\sigma_0 b^2}{S}$$

for plane strain. In this case

$$-\frac{\partial P}{\partial b} = -\frac{2P_L}{b}$$

and applying Eq. (3.20)

$$\begin{aligned} J &= -\frac{1}{B}\int_0^P \frac{\partial P}{\partial a} d\delta \\ &= \frac{1}{B}\int_0^P \frac{\partial P}{\partial b} d\delta \\ J &= \frac{2}{Bb}\int_0^\delta P_L \, d\delta, \end{aligned}$$

which agrees with Eq. (8.22).

8.3.2.2 Elastic Material

The accuracy of Eq. (8.22) for elastic materials was considered by Srawley [13]. It was found that for $S/W = 4$ and for $a/W > 0.5$ Eq. (8.22) is a good approximation.

8.3.2.3 Elastic-Plastic Material

The accuracy of Eq. 8.22 was explored experimentally by comparing J from Eq. (8.22) to the multiple specimen method [14]. It was found that for $S/W = 4$ that Eq. (8.22) is valid for $0.4 < a/W < 0.6$.

8.3.3 Engineering Approach to Elastic-Plastic Fracture Analysis

Assuming that the conditions for $J = J_C$ for crack initiation and $J(a + \Delta a) = J_R(\Delta a)$ for J controlled crack growth are satisfied and that valid material test data are available, the analysis of fracture in an application will depend principally on the determination of J and dJ/da. An engineering approach, based on an interpolation of elastic and fully plastic solutions and on handbook results can be used for such analyses [15]. In this method J, the load line displacement due to the crack, v_L^c, and the crack mouth opening displacement, v_C, can be interpolated as

$$\begin{aligned} J &= J_e + J_{pl}, \\ v_L^c &= v_{Le}^c + v_{Lpl}^c, \\ v_C &= v_{Ce} + v_{Cpl}, \end{aligned} \tag{8.23}$$

where the e and pl subscripts denote the elastic and plastic part of each quantity.

We know already that in the linear elastic case

$$J_e(a, P) = \frac{K_I^2(a, P)}{E} = P^2 f \quad (a,\ \text{geometry}), \tag{8.24}$$

where P is the load and a is the crack length. Under small scale yielding conditions the estimate of J can be improved by using the effective crack length concept, where a in the above is replaced by

$$a_{eff} = a + \phi r_y, \tag{8.25}$$

$$r_y = \frac{1}{\beta\pi}\left(\frac{n-1}{n+1}\right)\left(\frac{K_I(a)}{\sigma_0}\right)^2, \tag{8.26}$$

$$\phi = \frac{1}{1+(P/P_0)^2}. \tag{8.27}$$

For plane stress use $\beta = 2$. For plane strain use $\beta = 6$.

Under fully yielded conditions in a power-law hardening material, $\gamma \sim \alpha\sigma^n$, thus since $\sigma \sim P$ (by equilibrium), $\gamma \sim \alpha P^n$ and $J \sim \sigma \cdot \gamma \sim \alpha P^{n+1}$. Thus under fully yielded conditions the non-dimensional J can be expressed in non-dimensional form as

$$\frac{J_p}{\sigma_0\gamma_0 b} = \alpha\left(\frac{P}{P_0}\right)^{n+1} \hat{J}_p(a/W, n), \tag{8.28}$$

where P_0 is the limit load for a rigid plastic material with yield stress σ_0, W is the width of the component and b is the uncracked ligament length.

The value of J is then interpolated as

$$J = J_e(a_{eff}) + J_p(a). \tag{8.29}$$

For small terms the first term will dominate. The second term dominates in LSY. Similar expressions can be written for the crack mouth opening displacement and the load-point displacement, experimentally measurable quantities.

In reference [15] the plastic parts of J, v_L^c, v_C are given as

$$J_p = \alpha\sigma_0\gamma_0 b \cdot h_1(a/W, n)(P/P_0)^{n+1}, \tag{8.30}$$

$$v_{Lpl}^c = \alpha\sigma_0\gamma_0 b \cdot h_3(a/W, n)(P/P_0)^n, \tag{8.31}$$

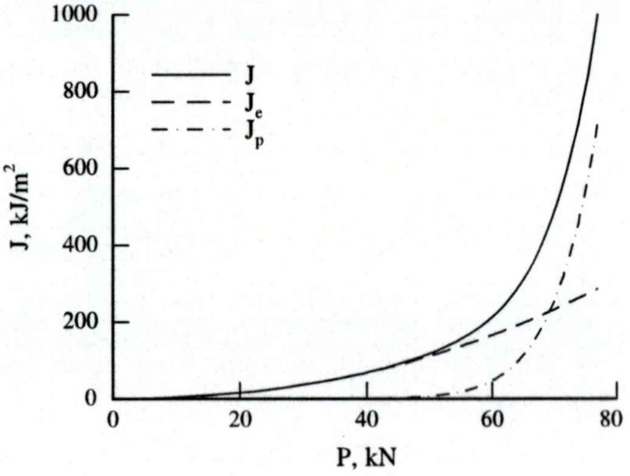

Fig. 8.11 J for three point bend specimen calculated using engineering approach in which J is calculated using the interpolation $J = J_e(a_{eff}) + J_p(a)$

$$v_{Cpl} = \alpha\sigma_0\gamma_0 b \cdot h_2(a/W, n)(P/P_0)^n, \tag{8.32}$$

where the fully plastic solutions $h_1, \ldots, h_3$ are tabulated in references [15, 16] for a number of geometries including test specimens and geometries such as pipes and other potential applications. The elastic parts of the interpolation, such as J_e, can be calculated using tabulated stress intensity factor handbook solutions [17, 18] or using energy or finite element methods.

8.3.3.1 Sample Calculation

Consider a three point bend specimen made of steel with $\sigma_0 = 700$ MPa, $n = 10$, $\alpha = 0.5$, $E = 200$ GPa. The dimensions (see Table 5.1 for definitions) are $B =$ 10 mm, $s = 300$ mm, $W = 75$ mm and $a = 28$ mm. This specimen is thin enough that it is appropriate to model it as plane stress. The plane stress limit load is given in [15] as $P_0 = 0.536\sigma_0(W - a)^2/(S/2) = 54{,}961$ N. The plane stress value of $h_1(a/W = 0.375, n = 10)$ is (from [15]) $h_1 = 0.307$. K_I is given by the equation in Table 5.1. In the calculation a_{eff} is calculated based on the actual crack length, a, then a_{eff} is substituted into the equation for J_e. For the three point bend specimen $b = W - a$.

Combining the results and using the equation for K_I

$$J = \frac{P^2 S^2 F^2(a_{eff}/W)}{B^2 W^3 E} + \alpha\sigma_0\gamma_0(W - a)h_1(a/W, 10)\left(\frac{P}{P_0}\right)^{n+1},$$

where F is given at the bottom of Table 5.1. The resulting J vs. P is shown in Fig. 8.11. Until P is greater than the limit load, the J_e contribution is much larger than J_p.

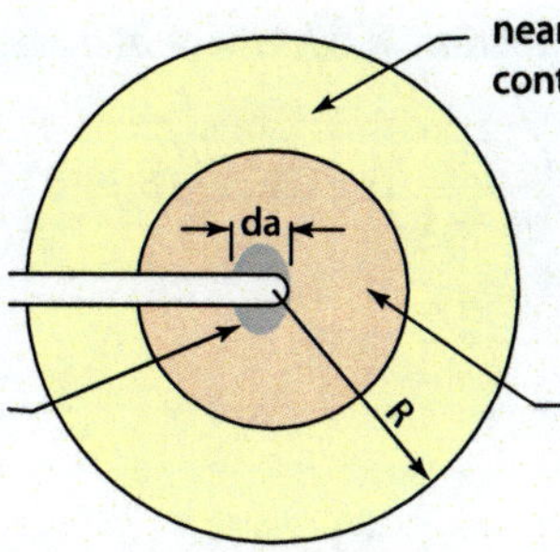

Fig. 8.12 Model for crack growth increment da under J controlled conditions

8.4 Fracture Criteria and Prediction

8.4.1 J Controlled Crack Growth and Stability

From the HRR field we know that under conditions of constrained loading J gives the amplitude of the crack tip stress and strain fields and of the crack tip opening displacement. Thus as long as full plane strain constraint exists a criterion for crack initiation is

$$J(a, P) \geq J_{IC}, \tag{8.33}$$

where $J(a, P)$ is the value of J calculated for a given crack length a and load, P.

However, as the crack grows the strain fields change considerably. In addition, plane strain constraint may not be maintained as the crack approaches a free edge. Thus we pose the question: Under what conditions does J control crack growth?

Consider an increment of crack growth, da from a stationary crack under ssy conditions. As shown in Fig. 8.12, in a region near the crack tip that scales with da there is a region of elastic unloading. In the annular region outside the elastic unloading zone there is non-proportional loading. Outside this zone is a zone of nearly proportional loading in which the stress and strain are determined by J. The size R of the outer zone is expected to scale with the crack length or other relevant in-plane dimension. Clearly one condition for J controlled crack growth is that $da \ll R$ so that the zone of elastic unloading is well embedded in the nearly proportional loading region.

From Eq. (7.63) the strain ahead of a stationary crack may be written as

$$\gamma_{ij} = k_n J^{\frac{n}{n+1}} r^{\frac{-n}{n+1}} \tilde{\gamma}_{ij}(\theta, n).$$

For an increment of crack growth, da and corresponding change in J of dJ (due to the J–R curve) the increment in strain is [19]

$$d\gamma_{ij} = \frac{\partial \gamma_{ij}}{\partial a} da + \frac{\partial \gamma_{ij}}{\partial J} dJ.$$

For crack growth in the x_1 direction $\partial(\cdot)/\partial a = -\partial(\cdot)/\partial x_1$. Thus $d\gamma_{ij}$ is

$$d\gamma_{ij} = k_n J^{\frac{n}{n+1}} \left(\frac{n}{n+1}\right) r^{-\frac{n}{n+1}} \left[\frac{da}{r}\tilde{\beta}_{ij} + \frac{n}{n+1}\frac{dJ}{J}\tilde{\gamma}_{ij}\right],$$

$$\tilde{\beta}_{ij} = \tilde{\gamma}_{ij}\cos\theta + \frac{\partial\tilde{\gamma}_{ij}}{\partial\theta}\sin\theta. \tag{8.34}$$

As $n \to \infty$

$$d\gamma_{ij} = \frac{J}{r\sigma_0}\left[\frac{da}{r}\tilde{\beta}_{ij} + \frac{dJ}{J}\tilde{\gamma}_{ij}\right]. \tag{8.35}$$

For J dominance the second term must dominate over the first, i.e.

$$\frac{dJ}{J} \gg \frac{da}{r}. \tag{8.36}$$

Assuming that the crack began to grow when $J = J_{IC}$, let us define a length scale D as

$$D \equiv \frac{J_{IC}}{dJ/da}. \tag{8.37}$$

The condition for J dominance may now be written as $r \gg D$, i.e. outside a region of size D the fields are controlled by J. To ensure a annular region of nearly proportional loading $D \ll r < R$.

Taken together the two conditions for J controlled crack growth are that $da \ll R$ and $D \ll R$. How is the region R to be determined in practical applications? Based on the comparison of the asymptotic and full elastic solutions we know that R will be a fraction of the uncracked ligament length, b or other relevant in-plane dimension. Finite element analyses of three SENB specimens examining the extent of crack growth needed for J to lose path independence suggest a bound on crack growth of [15]

$$da < 0.06b,$$

and a bound on D of

$$D < b/\omega,$$

where (for SENB) $\omega = 10$. In addition the plane-strain constraint condition $\rho \equiv b/(J_{IC}/\sigma_0) > 25$. For situations dominated by tension the conditions are more stringent, $\omega > 80$ and $\rho > 200$.

As a sample calculation consider the data of Joyce and Link [20] for a ductile steel tested with the SENB specimen with $a/W = 0.5$ and $b = 25$ mm. From their data, $J_{IC} = 0.20$ MPa·m and $dJ_R/da = 87$ MPa. Thus $D = 0.20/87.0 = 0.0023$ m $= 2.3$ mm. The ratio $\omega = c/D = 25/2.3 = 11$ is larger than 10 satisfying the condition for a J controlled annular region. The plane-strain constraint conditions ($\rho > 25$) were satisfied as well. The J–R data can be applied for da up to $da = 0.06 \cdot b = 0.06 \cdot 25 = 1.5$ mm.

Although 1.5 mm is a small amount of crack growth it is sufficient to assess the stability of fracture starting from a stationary crack by comparing the initial slope of the J_R curve to the slope of $J(a,P)$. The initial slope of the J resistance curve can expressed in a non-dimensional form as the "tearing modulus,"

$$T_0 \equiv \frac{E}{\sigma_0^2}\frac{dJ_R}{da}\bigg|_{\Delta a=0}. \tag{8.38}$$

Values of T_0 range from 0.1 to over 200 [19] with typical value on the order of 20–100 for ductile metals. As with elastic fracture, if the slope of the available J is less than the slope of the fracture resistance curve then fracture will be stable, i.e. $T < T_0$, where $T = (E/\sigma_0^2)(dJ(a,P)/da)$ and $J(a,P)$ is the value of J as a function of crack length, calculated based on the loading and geometry of the problem at hand.

As in the elastic fracture case, dJ/da depends on the stiffness of the loading as well as the geometry of the component being analyzed.

Consider a component loaded under fixed total displacement (v_T) conditions but through a compliant system with an effective compliance C_M as sketched in Fig. 4.6. Let v_T be the total displacement, which will be held constant, and v_L the load-line displacement (or equivalent displacement) of the component under consideration. To look at stability when v_T is fixed,

$$0 = dv_T = \frac{\partial v_L}{\partial P}\bigg|_a dP + \frac{\partial v_L}{\partial a}\bigg|_P da + C_M\,dP.$$

The change in J is

$$\frac{dJ}{da} = \frac{\partial J}{\partial a}\bigg|_P + \frac{\partial J}{\partial P}\bigg|_a \frac{dP}{da}.$$

Solving for dP, and substituting,

$$\frac{dJ}{da}\bigg|_{v_T} = \frac{\partial J}{\partial a}\bigg|_P - \frac{\partial J}{\partial P}\bigg|_a \frac{\partial\delta/\partial a_P}{[\partial v_L/\partial P|_P + C_M]}. \tag{8.39}$$

The above terms and hence stability of crack growth can be determined using finite element analysis, closed form solutions or using the EPRI estimation method where functional forms for δ and J in terms of the load, crack length and hardening have been tabulated and can be differentiated to determine T.

8.4.2 *J–Q Theory*

As discussed in Sect. 7.6.2 the constraint at the crack tip is lower for plane stress than for plane strain and is generally lower for large scale yielding than for small scale yielding. The maximum normal stress in an elastic-perfectly plastic material can vary from $1.15\sigma_0$ to $2.97\sigma_0$. The hydrostatic stress can vary from $1.8\sigma_0$ to $7.17\sigma_0$. It should be expected that such large differences in stress could lead to different mechanisms of fracture and to different fracture toughnesses. High tensile stresses would promote failure and transition to cleavage fracture at temperatures where the material may be prone to either ductile fracture by void formation and coalescence or to cleavage failure. In such cases a single parameter characterization of the crack tip fields and of the fracture toughness may not be adequate. Depending on the problem at hand and on whether the material is prone to ductile-cleavage transition ignoring the effect of different constraints between the laboratory test specimen and the application may be conservative or non-conservative to a degree that may not be determinable. Thus it is desirable to have a fracture model with additional parameters that can account for the differing constraint values.

Through a series of computational simulations it has been shown that to a good approximation the stress field ahead of a power law hardening material can be expressed as the HRR field with the addition of a hydrostatic term of magnitude $Q\sigma_0$ [21, 22] (Q here is not the same Q used to represent generalized loads). The resulting field can be written as

$$\sigma_{ij}(r,\theta) = \sigma_0 \left(\frac{J}{\alpha\gamma_0\sigma_0 I_n r} \right)^{1/(n+1)} \tilde{\sigma}_{ij}(\theta) + Q\sigma_0\delta_{ij}. \tag{8.40}$$

The dependence of Q on the load and geometry was determined through a series of finite element simulations. Small scale yielding with a varying T stress was simulated through a boundary layer analysis in which tractions from the K_I field plus a T stress are imposed as boundary conditions for a circular region containing a crack. The effects of finite geometry and large scale yielding were studied for a center cracked panel (CCP) loaded with stresses $\sigma_{22} = \sigma_\infty$ and $\sigma_{11} = \lambda\sigma_\infty$ and a single edge notched beam (SENB) loaded in pure bending as shown in Fig. 8.13. In these simulations (or in the analysis of an experiment or an application) Q is determined by taking the difference between the computed and HRR field hoop stresses, i.e.

$$Q \equiv \frac{\sigma_{\theta\theta} - (\sigma_{\theta\theta})_{HRR}}{\sigma_0} \quad \text{at } \theta = 0,\ r = 2J/\sigma_0. \tag{8.41}$$

The distance $r = 2J/\sigma_0$ is chosen to that Q is evaluated outside the finite strain region but within the region dominated by the J–Q field.

The results of the computations are shown in Fig. 8.14. In SSY decreasing the T stress decreases Q. This can be modeled (for SSY only) as

$$Q = c_0 + c_1\frac{T}{\sigma_0} + c_2\left(\frac{T}{\sigma_0}\right)^2, \tag{8.42}$$

where for $n = 5$, $c_0 = -0.1$, $c_1 = 0.72$, $c_2 = -0.42$ and for $n = 10$, $c_0 = -0.05$, $c_1 = 0.81$, $c_2 = -0.54$. In the CCP geometry, if a high stress parallel to the crack tip is maintained ($\lambda = 1.0$) then the constraint is maintained and $Q = 0$ for all load levels. However, in the absence of an external stress parallel to the crack, $\lambda = 0$,

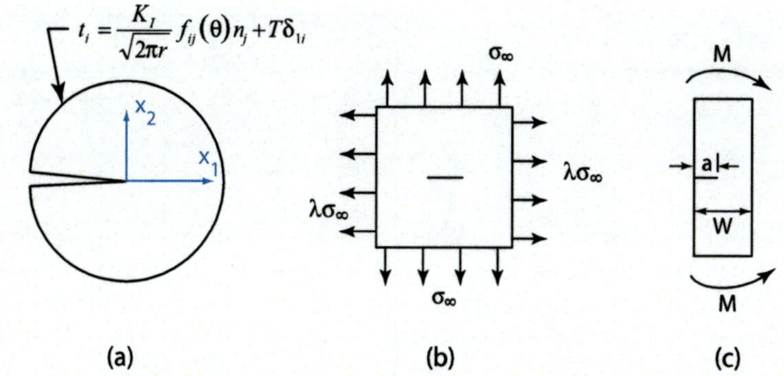

Fig. 8.13 Geometries for FEM computation of Q. (**a**) In SSY model the K_I field plus T stress is imposed as traction BC on a circular region around the crack tip. (**b**) A biaxial stress is applied to the center crack panel (CCP). (**c**) Single edge notched beam (SENB) in pure bending

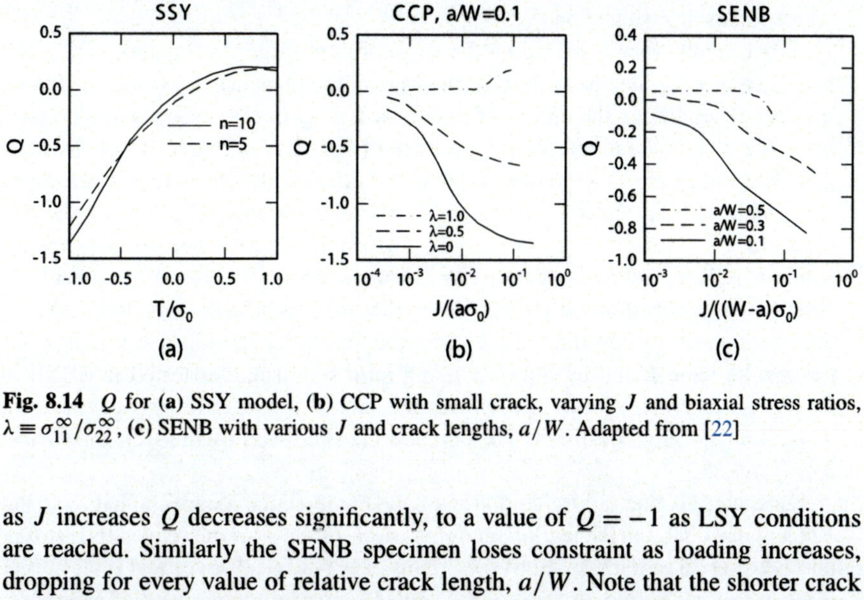

Fig. 8.14 Q for (**a**) SSY model, (**b**) CCP with small crack, varying J and biaxial stress ratios, $\lambda \equiv \sigma_{11}^{\infty}/\sigma_{22}^{\infty}$, (**c**) SENB with various J and crack lengths, a/W. Adapted from [22]

as J increases Q decreases significantly, to a value of $Q = -1$ as LSY conditions are reached. Similarly the SENB specimen loses constraint as loading increases, dropping for every value of relative crack length, a/W. Note that the shorter crack specimens ($a/W = 0.1$, and $a/W = 0.3$) lose constraint much more quickly than the deep crack specimens recommended by ASTM 1820 [5].

Applying the theory that cleavage fracture will occur when the maximum tensile stress at a critical distance ahead of the crack reaches a critical value (RKR theory) [23] the fracture toughness envelope $J_C(Q)$ was calculated for the SSY case to be

$$\frac{J_c}{J_c^{T=0}} = \left(\frac{\sigma_c/\sigma_0 - Q}{\sigma_c/\sigma_0 - Q^{T=0}} \right)^{n+1}. \tag{8.43}$$

This equation is plotted in Fig. 8.15 for $n = 5$ and $\sigma_c/\sigma_0 = 2, 3, 4$. A strong increase in the critical J value is seen as constraint (Q) decreases. The effect is more

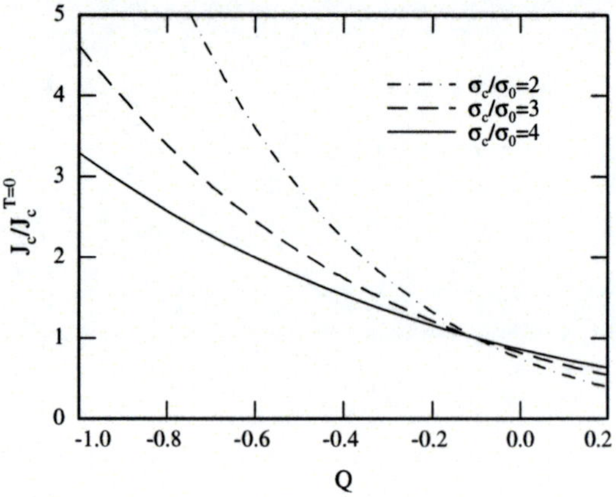

Fig. 8.15 J–Q envelope predicted by RKR theory [23] for a material with $n = 5$

pronounced for lower σ_c/σ_0 values. Furthermore the effect is more pronounced for larger values of n (lower hardening).

Experimental results regarding the applicability of the J–Q theory appear to be few. However for metals the fail by cleavage in the presence of plastic deformation the idea of predicting the onset of fracture by a J–Q envelope appears to be sound. By varying the a/b ratio in SENB specimen of 10, 25, 50 mm thicknesses Kirk et al. [24] determined the J–Q toughness locus for A515 Grade 70 steel at room temperature. This material failed by cleavage. The results, shown in Fig. 8.16(a) show that the value of J for cleavage increases by a factor of approximately 5 as the constraint is reduced (decreasing Q). This is true for all thicknesses of material tested. Note that in contrast to the typical notion that toughness decreases with increasing thickness, these results show the highest toughness for the thickest samples. Similar results were found by Sumpter using mild steel samples tested in SENB and CCP geometries at −50°C [25]. The results, shown in Fig. 8.16(b) show increasing J_c with decreasing constraint. The data are matched well by the JKR theory using $\sigma_c/\sigma_0 = 4.25$.

For materials that fail by void growth and coalescence it appears that J_c is independent of Q but that the tearing modulus (slope of the J–R curve) depends strongly on Q. Using 25 mm thick SENB specimens of HY-80 steel Joyce and Link [20] varied a/b to vary Q and performed J_{IC} and J_R tests following the ASTM standard. The results showed that J_c was independent of Q, but that the tearing modulus, shown in Fig. 8.17 increased by a factor of two with decreasing constraint.

8.4.3 Crack Tip Opening Displacement, Crack Tip Opening Angle

Two successful criteria for predicting stable crack growth under Mode-I loading in elastic-plastic materials are that the crack achieves a critical displacement (CTOD) at a specified distance behind the crack or the crack tip opening angle (CTOA)

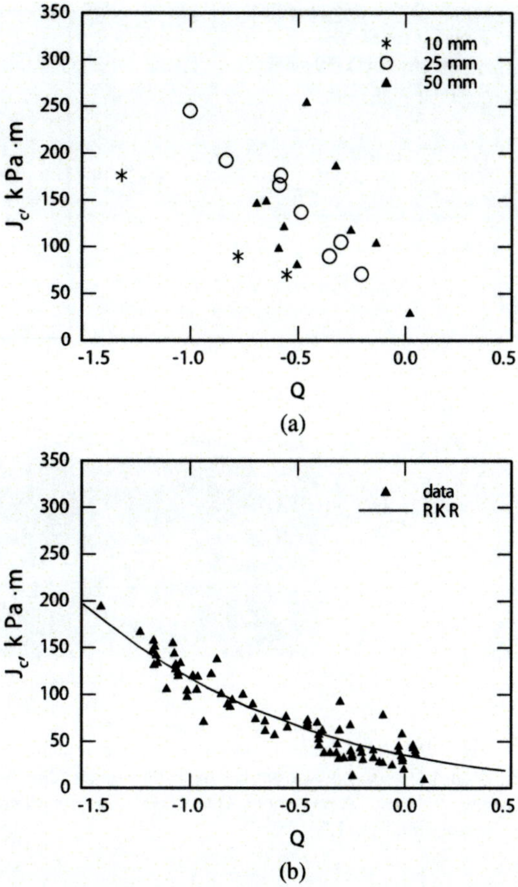

Fig. 8.16 J_c–Q loci for two steels. (**a**) A515 Grade 70 steel at 20°C, using three thicknesses of SENB specimens. Adapted from [24]. (**b**) Mild steel at −50°C using SENB and CCP specimens. RKR theory line for $\sigma_c/\sigma_0 = 4.25$. Adapted from [25] as reported in [26]

reaches a critical value, i.e. $\psi = \psi_c$ [27]. CTOD at a distance and CTOA are related by geometry, thus these two criteria are essentially equivalent.

Stable tearing of an aluminum alloy is shown in Fig. 8.18 along with lines of 6.2°, 6.8°, and 6.9° marking CTOA for various distances behind the crack tip. Although the measurement of CTOA is not without ambiguities, studies generally show that after a small amount of crack growth, typically on the order of one plate thickness, the CTOA is essentially constant as the data in Fig. 8.19 demonstrate for crack growth in compact tension and center crack tension panels of aluminum alloy 2.3 and 25 mm thick. Note the lower critical CTOA for the thicker specimen.

The application of the CTOA criterion relies on the use of calibrated finite element analyses. In the crack growth analysis, the load or displacement is incremented in an FEM computation until the critical CTOA is reached. At this point the load on the crack tip node is relaxed to zero in a series of equilibrium iterations, thus increasing the crack length by one node. The applied load is then increased and the process repeated.

Fig. 8.17 Tearing modulus, T_0 vs. Q for SENB specimens of HY-80 steel. Adapted from [20]

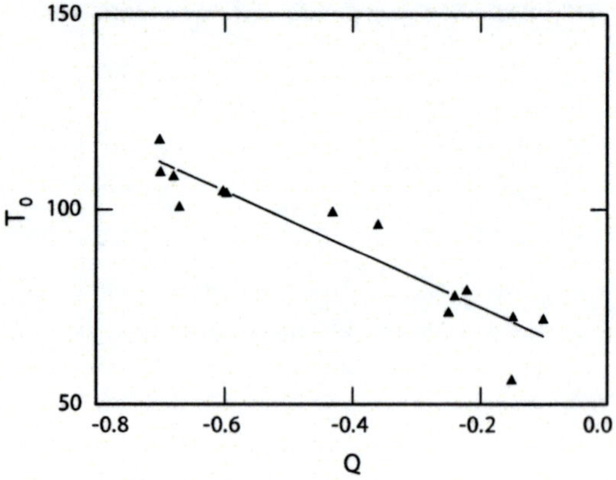

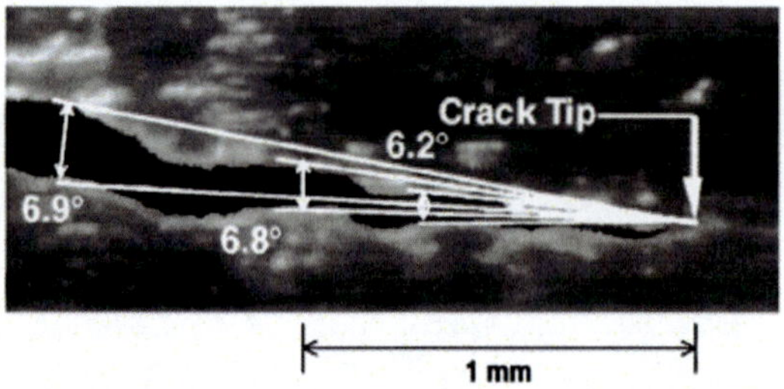

Fig. 8.18 Video image of crack opening during stable tearing of a thin sheet of an aluminum alloy. From [27]. Reprinted from Engineering Fracture Mechanics, Vol. 70, J.C. Newman, M.A. James and U. Zerbst, "A review of CTOA/CTOD fracture criterion," pp. 371–385, Copyright (2003), with permission from Elsevier

Sample results are shown in Fig. 8.20 where the applied load vs. crack extension is shown for experiments and simulations of center cracked tension sample of aluminum alloy. The FEM results show that plane strain is too stiff, overestimating the peak load, while plane stress underestimates the peak load. The 3D analysis and plane strain core analysis (not shown in the plot) match the experiments very closely. The analysis of the peak load is important since this determines the maximum loads that can be applied prior to the onset of unstable crack growth.

Both 2D and 3D analyses can be used for the simulations. Two-dimensional analyses will clearly be faster to perform and make more sense when dealing with fracture in plate and shell structures. However, studies show that both the plane stress and plane strain analyses using CTOA are not accurate due to insufficient constraint in the plane stress model and over constraint in the plane strain model. Thus the idea of the "plane-strain core" was developed. The plane-strain core is a layer of elements along the prospective crack line in which plane strain conditions are prescribed. Away from the crack line plane stress conditions are prescribed. This model provides sufficient near crack tip constraint while allowing plane stress deformation away from the crack.

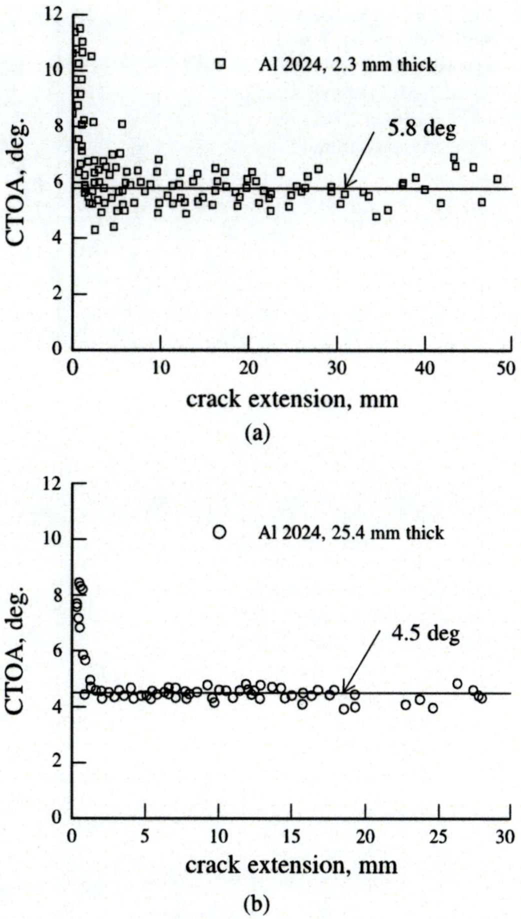

Fig. 8.19 (**a**) Crack tip opening angle measured for thin sheets (2.3 mm) of aluminum 2024 using a range of M(T) and C(T) geometries and (**b**) measured for a thick sheet (25.4 mm). CTOA is constant after a few mm of crack growth. Note reduced CTOA for thicker sample due to increased constraint. Adapted from [27]

Three dimensional analyses can also be performed and have been shown to be accurate. As with the plane strain core model the 3D model requires calibration of the critical CTOA and δ_c. An example calibration is shown in Fig. 8.21 where the failure load for several plate widths is plotted along with test data. In the calibration the parameters δ_c and ψ_c are iterated until the experimental and simulation data agree across a range of W values. Note that 3D analyses may require inordinate amounts of computer and modeling effort when applied to plate and shell structures.

8.4.4 Cohesive Zone Model

The Dugdale model discussed in Sect. 7.2 can be generalized to a model in which the stresses in the yield zone ahead of the crack are a function of the displacement

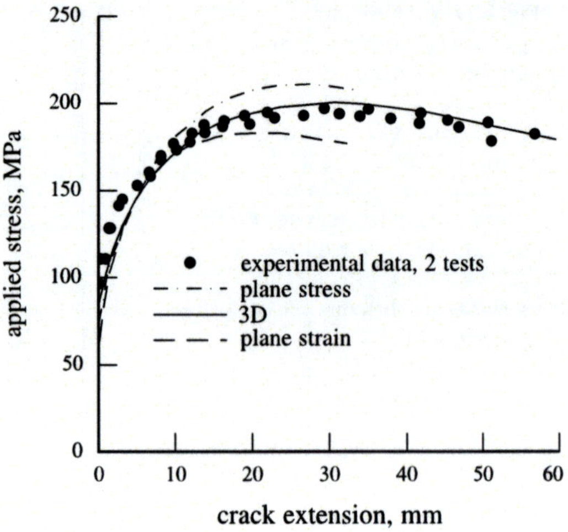

Fig. 8.20 Measured and simulated load vs. crack extension for 2.3 mm thick, Al 2024 M(T) test specimen using a critical CTOA of 5.25°. Adapted from [27]

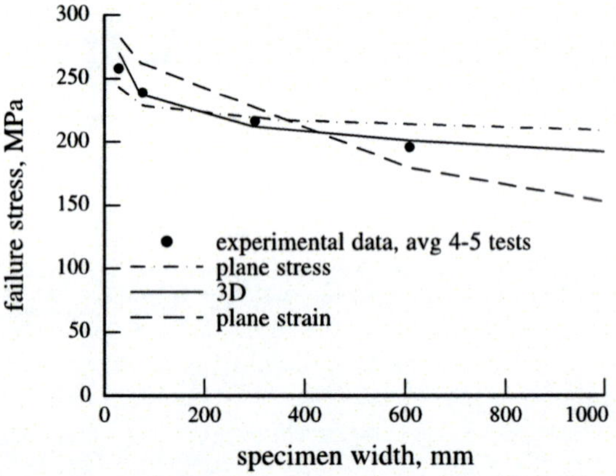

Fig. 8.21 Measured and calculated failure stresses for center cracked panels of 2024 aluminum alloy of different widths. Compares plane stress, plane strain and three dimensional simulation results. Adapted from [27]

across the yield zone, rather than a constant value. This is known as the "cohesive zone", or "cohesive forces" model and was pioneered by Barenblatt [28]. Here we will discuss the use of the cohesive zone model for Mode-I; the concept can be generalized to mixed-mode. See [29] for a review on cohesive zone modeling.[1]

A schematic normal traction-separation law is shown in Fig. 8.22. At the tip of the cohesive zone the opening is zero. Assuming that the crack is loaded to the point of incipient fracture at the back of the cohesive zone, the opening is equal to δ_C and the stress is zero. The most important parameters in the cohesive zone model are the

[1]The cohesive zone model is one of the most popular approaches in computational fracture mechanics. A January 2008 keyword search among journal articles for "cohesive zone" produced over 1400 hits. The same search, repeated in July 2011, produced over 2200 hits.

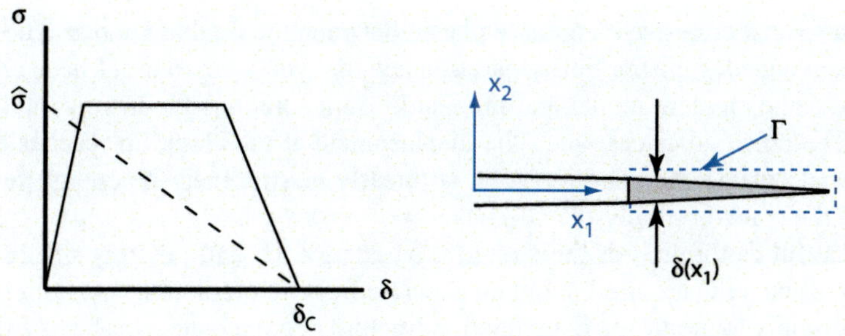

Fig. 8.22 Schematic of cohesive zone model. A nonlinear traction separation law is postulated to exist ahead of the crack. Shown are but two examples of the traction-separation laws. In the case of tensile cracks the traction is $\sigma = \sigma_{22}$ ahead of the crack and the separation, δ is $\delta \equiv u_2^+ - u_2^-$, i.e. opening displacement across the cohesive zone. In the concrete and rock fracture literature this is known as the "tension softening diagram"

strength, $\hat{\sigma}$, and the energy, Γ_0,

$$\Gamma_0 = \int_0^{\delta_C} \sigma(\delta)\, d\delta, \tag{8.44}$$

where $\sigma(\delta)$ is the normal stress (σ_{22}) across the cohesive zone.

Let us calculate the J integral for the cohesive zone model. The contour Γ is taken to lie along the cohesive zone as shown in Fig. 8.22. On this contour, $n_1 = 0$, hence J reduces to

$$J = -\int_\Gamma t_i \frac{\partial u_i}{\partial x_1}\, d\Gamma = -\int_{\text{c.z.}} \sigma_{22} \frac{\partial}{\partial x_1}(u_2^+ - u_2^-)\, dx_1$$
$$= -\int_{\text{c.z.}} \sigma(\delta) \frac{\partial \delta}{\partial x_1}\, dx_1 = -\int_{\text{c.z.}} \frac{\partial}{\partial x_1}\left[\int_0^{\delta} \sigma(\delta)\, d\delta\right] dx_1,$$
$$J = \int_0^{\delta_T} \sigma(\delta)\, d\delta,$$

where δ_T is the crack opening at the crack tip (back of the cohesive zone). If loaded to incipient fracture, then $\delta_T = \delta_C$ and

$$J = \int_0^{\delta_C} \sigma(\delta)\, d\delta = \Gamma_0. \tag{8.45}$$

Thus it is seen that the area under the $\sigma(\delta)$ curve is the energy required to propagate the crack.

8.4.4.1 Cohesive Zone Embedded in Elastic Material

As a fracture criterion the cohesive zone model is generally used in the context of a finite element analysis. Nodes along the prospective fracture path are doubled and a special cohesive element is placed between the doubled nodes. The cohesive element encodes the traction-separation law that you want to use. Loads are applied to the finite element model and the equilibrium solutions are incrementally solved for. The crack advances when the displacement at the crack tip reaches δ_C. If the material outside the cohesive zone is linearly elastic, then the energy required to drive the crack is simply $G = \Gamma_0$.

Careful calibration of the strength, $\hat{\sigma}$, energy, Γ_0 and possibly of the shape of the $\sigma(\delta)$ curve is required to obtain accurate fracture predictions. An approach used successfully to model the delamination fracture of composite facesheets from a honeycomb core was to determine $\hat{\sigma}$ using a "facewise tension test", an experiment in which the facesheet is pulled off of the core in pure tension. The fracture energy, Γ_0 was determined using a modification of the DCB test shown in Fig. 6.4 [30]. To verify the model parameters and the finite element procedures, the facesheet was modeled using plate elements, the core using continuum elements and the adhesive between the core and facesheet using cohesive elements. The DCB tests were simulated using the cohesive zone model and the resulting force versus crack length results were compared directly to experimental data. The model was applied to simulate a honeycomb core panel containing a pre-existing delamination under compressive loading. The pre-delamination first buckles, causing tension at the interface of the facesheet and core. Upon further loading the facesheet debonds from the core. Results from the finite element simulation agreed closely with experimental data.

8.4.4.2 Cohesive Zone Embedded in Elastic-Plastic Material

When the cohesive model is embedded in an elastic-plastic material, then the fracture energy flux will consist of the energy release rate at the crack tip plus the dissipation in the plastic zone around the crack tip. We showed in Sect. 8.1.3 that $G_{tip} = 0$ and that the energy required for elastic-plastic crack growth is the energy of plastic dissipation in the plastic zone around the crack tip. Of course this result is an approximation. Consider the case in which the energy needed to separate the atoms at the crack tip is zero, then the material would fracture with zero load and there would be no dissipation of energy in the plastic zone. In the cohesive zone model, $G_{tip} = \Gamma_0$ and the total fracture energy is $G = \Gamma_0 +$ (dissipation of energy in plastic zone).

Typically, Γ_0, which can be thought of as the energy absorbed by the failure processes local to the crack tip, is dwarfed in comparison with the energy dissipated by plastic deformation. Nonetheless, the resistance to crack growth is extremely sensitive to the details of the local cohesive description in the sense that a small increase in cohesive strength can lead to very large increases in toughness. It is analogous to a transistor (albeit a highly nonlinear one). In a transistor, the small base current controls the much larger collector current. In fracture mechanics, the small cohesive energy and strength of the atomic bonds control the fracture energy by dictating the extent of plastic deformation at the crack tip.

Finite element computations in which a cohesive zone model is embedded in an elastic-plastic power law hardening material show that K_{ss}, the K_I value required

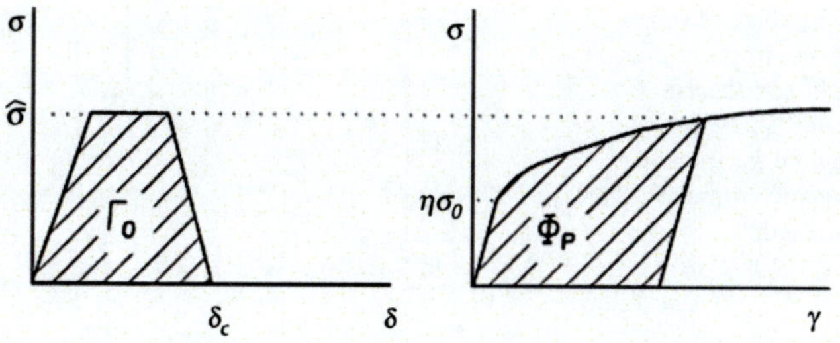

Fig. 8.23 The energy dissipated during crack growth is modeled as the sum of the cohesive energy, Γ_0 and the work of plastic deformation, $r_p\Phi_p$. From [32]

for steady state crack propagation, increases rapidly with the ratio of $\hat{\sigma}/\sigma_0$, where σ_0 is the yield stress of the material [31], demonstrating the sensitivity of the fracture energy to the strength of the cohesive zone model.

A simple scaling model of this process can be constructed. Deformation in the plastic zone is modeled as a nonlinear spring with stress-strain relation

$$\frac{\gamma}{\gamma_0} = \frac{\sigma}{\sigma_0} + \left(\frac{\sigma}{\eta\sigma_0}\right)^n,$$

where the factor η accounts for the elevation of tensile stress in the plastic zone due to constraint. The fracture energy is the sum of the cohesive energy plus the dissipation of plastic energy, Φ_p, the shaded region in Fig. 8.23,

$$G_{ss} = \Gamma_0, \quad \hat{\sigma} \le \eta\sigma_0,$$

$$G_{ss} = \Gamma_0 + r_p\Phi_p, \quad \hat{\sigma} \ge \eta\sigma_0,$$

where r_p is the plastic zone size, given by the Irwin plastic zone estimate

$$r_p = \frac{E\Gamma_0}{\beta\pi\sigma_0^2},$$

where $\beta = 1$ for plane stress and $\beta = 3(1-\nu^2)$ for plane strain. A straightforward calculation shows that

$$\Phi_p = \sigma_0\gamma_0 \frac{\hat{\sigma}}{(1/n+1)\sigma_0}\left(\frac{\sigma}{\eta\sigma_0}\right)^n.$$

Putting the above together, the steady state crack growth resistance is estimated as

$$\frac{G_{ss}}{\Gamma_0} = 1 + \frac{1}{\beta\pi}\frac{\hat{\sigma}}{(1/n+1)\sigma_0}\left(\frac{\sigma}{\eta\sigma_0}\right)^n. \tag{8.46}$$

Results from Eq. (8.46) are plotted in Fig. 8.24 for plane stress assuming $\eta = 1.2$ (the crack tip opening stress in a non-hardening material is $\sigma_{22} = 1.2\sigma_0$).

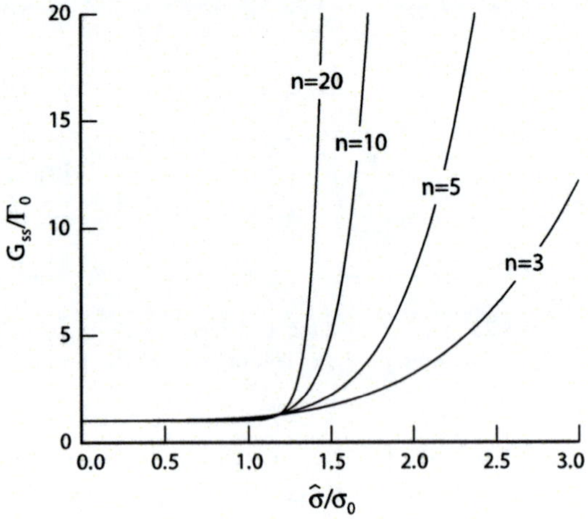

Fig. 8.24 Estimate of steady state toughness in plane stress using simplified cohesive model. Results show that energy required for fracture increases rapidly with the cohesive strength

References

1. B. Moran, C.F. Shih, Eng. Fract. Mech. **27**, 615 (1987)
2. B. Moran, C.F. Shih, Int. J. Fract. **35**, 295 (1987)
3. L.B. Freund, J.W. Hutchinson, J. Mech. Phys. Solids **33**, 169 (1985)
4. K.S. Bhalla, A.T. Zehnder, X. Han, Eng. Fract. Mech. **70**, 2439 (2003)
5. ASTM, *ASTM E 1820: Standard Test Method for Measurement of Fracture Toughness* (ASTM International, 2005)
6. W.R. Andrews, C.F. Shih, Elast.-Plast. Fract. **668**, 426–450 (1979)
7. D. Carka, C.M. Landis, Journal of Applied Mechanics (2010, in press). doi:10.1115/1.4001748
8. R. Barsoum, Int. J. Numer. Methods Eng. **11**, 85 (1977)
9. C.F. Shih, *Elastic-Plastic Fracture*, vol. 560 (1974), pp. 187–210
10. M. Stern, Int. J. Numer. Methods Eng. **14**(3), 409 (1979)
11. N. Levy, P.V. Marcal, W.J. Ostergren, J.R. Rice, Int. J. Fract. Mech. **7**, 143 (1971)
12. A.P. Green, B.B. Hundy, J. Mech. Phys. Solids **4**, 128 (1956)
13. J.E. Srawley, Int. J. Fract. **12**, 470 (1976)
14. P.M.S.T. de Castro, P. Spurrier, J. Hancock, Int. J. Fract. **17**, 83 (1984)
15. V. Kumar, M.D. German, C.F. Shih, Engineering approach for elastic-plastic fracture analysis. Tech. Rep. NP-1931, Electric Power Research Institute (1981)
16. A. Zahoor, *Ductile Fracture Handbook* (Electric Power Research Institute, Palo Alto, 1989)
17. H. Tada, P.C. Paris, G.R. Irwin, *The Stress Analysis of Cracks Handbook* (ASME Press, New York, 2000)
18. Y. Murakami, S. Aoki, *Stress Intensity Factors Handbook* (Pergamon, Oxford, 1987)
19. J.W. Hutchinson, *A Course on Nonlinear Fracture Mechanics* (Department of Solid Mechanics, The Technical University of Denmark, 1979)
20. J.A. Joyce, R.E. Link, Eng. Fract. Mech. **57**, 431 (1997)
21. N.P. O'Dowd, C.F. Shih, J. Mech. Phys. Solids **39**, 989 (1991)
22. N.P. O'Dowd, C.F. Shih, J. Mech. Phys. Solids **40**, 939 (1992)
23. R.O. Ritchie, J.F. Knott, J.R. Rice, J. Mech. Phys. Solids **21**, 395 (1973)
24. M.T. Kirk, K.C. Koppenhoefer, C.F. Shih, in *Constraint Effects in Fracture*, ed. by E.M. Hackett, K.H. Schwalbe, R.H. Dodds. Elastic-Plastic Fracture, vol. 1171 (ASTM International, Philadelphia, 1993), pp. 79–103
25. J.D.G. Sumpter, A.T. Forbes, in *Proceedings TWI/EWI/IS International Conference on Shallow Crack Fracture Mechanics Test and Applications* (Cambridge, UK, 1992)

26. N.P. O'Dowd, Eng. Fract. Mech. **52**, 445 (1995)
27. J.C. Newman, M.A. James, U. Zerbst, Eng. Fract. Mech. **70**, 371 (2003)
28. G.I. Barenblatt, Adv. Appl. Mech. **7**, 55 (1962)
29. M. Elices, G.V. Guinea, J. Gomez, J. Planas, Eng. Fract. Mech. **69**(2), 137 (2002)
30. T.S. Han, A. Ural, C.S. Chen, A.T. Zehnder, A.R. Ingraffea, S. Billington, Int. J. Fract. **115**, 101 (2002)
31. V. Tvergaard, J.W. Hutchinson, J. Mech. Phys. Solids **40**, 1377 (1992)
32. A.T. Zehnder, C.Y. Hui, Scr. Mater. **42**, 1001 (2000)

Index

A.T. Zehnder, *Fracture Mechanics*,
Lecture Notes in Applied and Computational Mechanics 62,
DOI 10.1007/978-94-007-2595-9,